GM Crops
and the
Global Divide

Jennifer Thomson

PUBLISHING

A catalogue record for this book is available from the National Library of Australia.

ISBN: 9781486312658 (pbk)
ISBN: 9781486312665 (epdf)
ISBN: 9781486312672 (epub)

Published in print in Australia and New Zealand, and in all other formats throughout the world, by CSIRO Publishing.

CSIRO Publishing
Locked Bag 10
Clayton South VIC 3169
Australia

Telephone: +61 3 9545 8400
Email: publishing.sales@csiro.au
Website: www.publish.csiro.au

A catalogue record for this book is available from the British Library, London, UK.

Published in print only, throughout the world (except in Australia and New Zealand), by CABI.

ISBN 9781789248401

CABI
Nosworthy Way
Wallingford
Oxfordshire OX10 8DE
UK

CABI
We Work
One Lincoln Street, 24th Floor
Boston, MA 02111
USA

Tel: +44 (0)1491 832111
Fax: +44 (0)1491 833508
Email: info@cabi.org
Website: www.cabi.org

Tel: +1 (617)682-9015
E-mail: cabi-nao@cabi.org

Front cover image from Pixabay.

Edited by Kerry Brown
Cover design by James Kelly
Typeset by Desktop Concepts Pty Ltd, Melbourne
Index by Bruce Gillespie
Printed in Australia by McPherson's Printing Group

The paper this book is printed on is in accordance with the standards of the Forest Stewardship Council® and other controlled material. The FSC® promotes environmentally responsible, socially beneficial and economically viable management of the world's forests.

MIX
Paper from
responsible sources
FSC® C001695

Foreword

Do you care about your food, that of others in poor or rich countries, or about the agricultural environment and those who produce food? If so, then this book is for you. It tells about the scientific advances that led to improved, genetically modified crops (GMOs) and analyses the consequential debates about their safety, value and desirability since their introduction in the early 1990s. As the author teaches us, there is much to be learnt from the successes and failures concerning how GMOs have and have not been adopted around the world.

The author has been amidst the science of GMOs from the beginning, as have I, and knows the facts. However, unlike many who have written about GMOs, she lives in Africa, cares about what African countries, farmers and consumers need, and has created GMOs that overcome important deficiencies in African crops. Her research has focused on developing maize resistant to endemics of maize streak virus and to drought, both important issues in African agriculture.

While the book outlines the breakthroughs that led to the making of GMOs, its overall thrust is about recognising the need for the benefits of GMOs to be available and adopted in many of the disadvantaged parts of the world, including Africa. Most importantly, it reveals why these countries have been denied them or failed to adopt them to date, even those generated in Africa for Africa.

GMOs are over 25 years old. There is now a huge amount of evidence to show that today's GMOs have built-in improvements that aid farming, the health of farm workers, big and small farm economics, and the care of soils and environments in addition to the generation of higher yields. These may seem irrelevant to some consumers and retailers in the rich countries of the world who advertise 'GMO free' foods to boost their sales, but these GMO attributes can (and do) provide real benefits for those whose agriculture is on a knife edge because of ill-suited crops, bad soils and/or where poverty prevents investment into the crop production systems typical of richer countries.

The author is well qualified to write on these issues because she has been Chair and member of the South African Genetic Engineering Committee, is a member (and first Chair) of the board of the African Agricultural Technology Foundation (AATF), Vice-Chair of ISAAA (International Service for the Acquisition of AgriBiotech Applications), a Fellow of the Royal Society of South Africa and a former Vice-President of the Academy of Science of South Africa.

Much of the controversy over GMOs has arisen because there are many myths surrounding what GMOs are and what effects they have when eaten or grown. It is hard for us to know what is fact and what is fiction, what to believe and what to reject, given the debates. The author has used her extensive experience to explain the facts and misleading science, with supporting evidence, as well as to document some of the many myths deliberately used to prevent or hurt the adoption of GMO products in rich and poor countries. The book is worth reading just to enable us to be better informed on the myths and the damage they have done for some.

Then there are the risks. In life there are real risks and imaginary ones. It is easy to confuse them, especially when the facts are misunderstood. As a member of a rich Western society, I can afford to avoid would-be risks and not bother to assess them properly. But why should I presume to know what is a risk for crops and farmers in other societies and environments? Sadly, activists in food-rich Europe and the consequential European Union's legislation on GMOs built upon misplaced fear and/or personal preferences have had major effects, on Africa in particular, and stimulated a lack of confidence in the crops as food and as cash crops for export.

GMOs provoke the question of the place of science in crop improvement. Few of our crops are taken directly from nature. Most are human-made, built by spotting new individual plants with better characteristics and creating improved ones that combine the best of the new with the best of the old. We must be extraordinarily grateful to our scientific forefathers, going back thousands of years in many cases, who made the crop innovations we now rely on. They turned wild weeds into crops that kept them and us alive! Where would we be without the new food crops they developed?

Science is developed to gain knowledge and create opportunities for better lives and relief from hardship. In the 1960s almost no one had taken DNA from inside plants to study its information. Today we know most of the information within the DNA of our major crops and know how to interpret much of it. There is therefore certainty that we will be able to program plants to do a better job of serving humankind, from farmers to consumers. This is the story of the past and will be the story of the future. We need to learn from the GMO story, as detailed in this book, and how to develop trust in new, useful properties of plants.

Nostalgia for the crops of the past leaves agriculture, farmers and consumers stuck with yesterday's problems. All too often, the varieties of the past fail to cope with today's diseases, droughts, climate change, insect damage and poverty. Instead they perpetuate poverty, stress and ill health. If we fail to encourage our scientists to create beneficial changes or fail to embrace the new worthwhile opportunities, we are condemning others, present and future, to continued hardship.

This book challenges all to consider new scientific advances as they emerge and embrace them where appropriate to help overcome the hardships of today, and especially climate changes and the need to use less land and fewer inputs to feed ourselves. As this book reminds us: 'To have maintained global food production levels at 2016 levels, without GMOs, would have required farmers to plant an additional 10.8 million hectares (ha) of soybeans, 8.2 million ha of maize, 2.9 million ha of cotton and 0.5 million ha of canola, an area equivalent to the combined land area of Bangladesh and Sri Lanka.' May these land savings and higher yields come fully to all who want the benefits unhindered by myths, fake science, prejudice or ignorance. This book helps us to understand the roadblocks built in the rich countries that create the global divide.

Richard B Flavell, DSc FRBS CBE FRS
London, UK

Contents

Preface

The original title of this book was *GM Crops: The West versus the Rest*. However, the more I wrote the clearer it became that the world cannot be neatly divided into two such parts. Nor can it be split into developed and developing countries. For instance, if a country such as Canada, which would conventionally be labelled a developed country, is not constantly developing then surely it is regressing? So although it won't raise your eyebrows, the current title does state clearly what my book is all about.

One way in which countries are classified as developed or developing is on the basis of their GDP per capita. A somewhat arbitrary cut-off point that is informally recognised is US$12 000, although some economists prefer US$25 000. Perhaps a more realistic measurement would be the human development index (HDI), which includes life expectancy and levels of education as well as income. The difference becomes clear when we look at the case of China. Using GDP it is a developed country, but using HDI changes it to a developing one (Investopedia 2019).

Another way to divide countries has been suggested by Hans Rosling in his book *Factfulness* (Rosling 2018). He uses four divisions, levels 1–4, based on income as follows.

Level 1 inhabitants earn US$1 per day. They collect water from a distant waterhole, eat the same meal three times a day from food they grow themselves. There are about 1 billion people living in such dire poverty.

Level 2 people earn US$4 a day. They can buy food, their children go to school and one of them might land a job in a local industry. About 3 billion people live like this today.

Level 3 means that they are getting out of poverty. They work multiple jobs and can earn US$16 per day. They have electricity and their children may finish high school. There are about 2 billion people living like this.

Level 4 people earn more than US$32 per day, having had 12 years of education. The difficulty for the approximately 1 billion people at this level is that they have always known this level of income and find it difficult to understand the huge differences between the other three levels. Most of you who are reading this book belong in level 4.

According to Rosling, 'just 200 years ago, 85% of the world population was still on Level 1, in extreme poverty. Today the vast majority of people are spread out in the middle across Levels 2 and 3.'

It would therefore be a good idea, while reading this book, to bear these levels in mind and interpret the words 'developing', 'developed', 'the West' and 'the Rest' not as closed and defined categories, but as gradations from level 1 to 4.

Having explained all this, why did I write this book? I became aware of the tensions between certain level 4 people and their organisations and those in levels 1–3 regarding genetically modified (GM) crops when I was writing my last two books. In *Seeds for the Future: The Impact of Genetically Modified Crops on the Environment* (Thomson 2007) and *Food for Africa: The Life and Work of a Scientist in GM Crops* (Thomson 2013), I wrote about these issues. In *Seeds for the Future* I came across it starkly when writing about biosafety regulations, trade and legal issues in the European Union. I wrote: 'The EU maintains a precautionary approach and tends to say 'no' or 'let's wait', rather than 'yes'.' In *Food for Africa*, I wrote in the final chapter: '… the bridge across the agricultural genetic divide between African countries and those in the developed world needs to be crossed. This divide separates the genetically improved varieties … available in the developed world from those being used by resource-poor farmers in Africa.'

These comments led me to write another book about these divides and whether bridges could be formed to overcome them. I am an eternal optimist (otherwise I wouldn't still be living in South Africa!) and I therefore believe that science will beat prejudice.

References

Investopedia (2019) Top 25 developed and developing countries. Investopedia, 21 November 2019. <https://www.investopedia.com/updates/top-developing-countries/>

Rosling H (2018) *Factfulness*. Hodder and Stoughton, London.

Thomson JA (2007) *Seeds for the Future: The Impact of Genetically Modified Crops on the Environment*. CSIRO Publishing, Australia.

Thomson JA (2013) *Food for Africa: The Life and Work of a Scientist in GM Crops*. UCT Press, South Africa.

Acknowledgements

This is the first book I have written for which I have had an editor from day 1 and what a difference it has made! Lauren Webb, from CSIRO Publishing, has made writing this book significantly easier by commenting extremely cogently on every chapter within days of my submitting it. Coming to the topic with an outsider's viewpoint, she would write comments such as, '… for Chapter 7: countries that got it right and why, what do you mean by "got it right"?' And there I was assuming that everyone would automatically know what I had in mind.

Another excellent suggestion she made early on was that I should send every chapter to experts in the field. And I have had experts second to none who have spent a great deal of time and effort into giving me superb feedback and ideas for new topics that I should include. So, let me say my hugely grateful thanks to Dave Woods, my PhD supervisor and former Vice-Chancellor of Rhodes University, for Chapter 1 in which he also features. Likewise, Dick (Richard) Flavell, former Director of the John Innes Institute in the UK, who appears in Chapter 2 and made sure I had my facts and chronology of events correct. He also read the whole book and wrote the foreword. Thank you Dick, you are a real star! In addition, Doug McKalip from the US Department of Agriculture and Neil Hoffman from the Animal and Plant Health Inspection Service (APHIS) put me right on several issues. Justus Wesseler also appears in, and helped me enormously with, Chapter 3.

For Chapter 4 I am indebted to two colleagues from Kenya: Leena Tripathi from International Institute of Tropical Agriculture (IITA) and Sylvester Oikoh from the African Agricultural Technology Foundation (AATF), both of whose work is covered. Then came my economist and political scientist friends, Graham Brookes, director of the PG Economics, and Rob Paarlberg, from Harvard University, whose expertise was invaluable.

For the chapter on myths and communication, there were Kevin Folta and Craig Cormick, both great communicators themselves, Kevin from the USA and Craig from Australia. In addition, there were Mark Lynas (freelance journalist par excellence; if you haven't read his *Seeds of Science* do yourself a favour and do so), plus Sarah Evanega who runs the Alliance for Science at Cornell University, many of whose fellows also gave valuable input.

For Chapter 7 I need to thank Hennie Groenewald and Michael Gastrow from South Africa, Stuart Smyth from Canada and Martin Lema from Argentina. However, my biggest thanks have to go to Ron Herring from Cornell University. He gave incredibly helpful advice and sent me articles I had not come across for chapters 6, 7 and 10. The time you spent on this book, Ron, fills me with respect and gratitude.

Hennie Groenewald, from BioSafety SA, also gave input into Chapter 8, as did Ben Durham from the South African Department of Science and Innovation – many thanks. And then, for Chapter 10, besides Ron I am grateful to Mark Lynas. However, a special word of thanks goes to my young 14-year-old friend, Sophie, who brought the book *Factfulness* to my attention and pointed out that the terms 'West', 'Rest', 'developed' and 'developing' are gross oversimplifications.

And finally, my publisher from CSIRO Publishing, Eloise Moir-Ford, who took over from Lauren and helped me admirably with the final hurdles. With so many people to thank, the reader might well wonder whether, in fact, I actually wrote the book. Perhaps I should just say that I couldn't have done it without them all.

1
Genetically modified organisms make their entrance

I first became aware of the scientific advances that would allow researchers to genetically modify organisms in August 1974 while on honeymoon in Sweden. My PhD supervisor, David Woods, who was on sabbatical leave in Norway, had, together with his family, joined my husband and me while we were staying on an island off the coast of Stockholm. 'This could change the way we microbiologists carry out our research', he told me. 'It could also revolutionise medical research with life-saving drugs being made in micro-organisms such as bacteria and yeast.'

As a result, I was not totally ignorant of this new field of research when I arrived at Harvard Medical School in Boston in September 1974 to begin my postdoctoral fellowship. The Department of Microbiology and Immunology was abuzz with the new recombinant DNA research technology that enabled scientists to cut and splice together DNA from different organisms, recombining the DNA in totally novel ways and giving them new characteristics or traits. Researchers could also use this technology to insert a gene they were interested in learning about into a bacterium to study its properties and functions.

For instance, Nagata *et al.* (1980) cloned a fragment of DNA from interferon-producing human white blood cells and showed that it had biological interferon activity. The word 'clone' simply means making an identical copy of something. In this case a fragment of DNA was inserted into a plasmid vector and introduced into the bacterium *Escherichia coli*. The researchers were then able to study interferon activity in *E. coli*, an organism much easier to work with than human white blood cells. The head of the laboratory responsible for this and further work on interferon, Charles Weismann, had already by this time, cofounded the biotechnology company, Biogen.

Another example came from the laboratory of Roy Curtiss (of whom more later). They were interested in the bacterium *Streptococcus mutans*, the main cause of dental caries and thus likely to be one of the most ubiquitous infectious agents worldwide. At the time, *S. mutans* was difficult to analyse genetically by classical methods and thus, by cloning several its genes into *E. coli*, the protein products could be analysed. In this way it was possible to determine the genes that are responsible for the ability of *S. mutans* to colonise the oral cavity and cause tooth decay (Curtiss *et al.* 1983).

Another early example was the cloning and characterisation of the vitellogenin structural gene of *Xenopus laevis*, the African clawed frog, widely used in research as a model system for humans. The most famous use of the frog was as a test for pregnancy developed by scientists at the University of Cape Town in South Africa. The test was done by injecting a frog with a woman's urine and was widely used from the 1930s until the 1960s (Shapiro and Zwarenstein 1934). The vitellogenin protein is the main egg storage protein precursor and is important in the development of many animals that give rise to their offspring via eggs.

A further example enabled scientists to determine that the mouse gene coding for dihydrofolate reductase rendered the *E. coli* host cells, into which it was cloned, resistant to the drug trimethoprim, which is used mainly for the treatment of bladder infections (Chang *et al.* 1978). The scientist in whose Stanford University laboratory this work was carried out was Stanley Cohen, who later teamed up with Herb Boyer to form the first biotechnology company, Genentech. As with Biogen, this will be discussed later in this chapter.

And then, of course, came the cloning of genes for pharmaceutical purposes, such as human growth hormone (Goeddel *et al.* 1979a), human insulin (Goeddel *et al.* 1979b), urokinase, used to treat blood clots (Ratzkin *et al.* 1981), and somatostatin, which regulates several human hormones (Itakura *et al.* 1977). However, those applications will be dealt with in detail further in this chapter.

In retrospect, however, genetically modified organisms (GMOs) appeared on the scene in a rather dramatic way. Paul Berg, a biochemist at Stanford, was one of the first scientists to develop recombinant DNA technology, or rDNA as it soon became known. Berg's research applied knowledge and techniques developed in the 1950s and 60s by earlier scientists. In particular, his work relied on the use of restriction enzymes, which Werner Arber had discovered in bacteria in the 1960s (Arber 1974). These enzymes provide the bacteria with a defence mechanism against invading viruses. Once the virus has injected its DNA into the bacterium, the enzymes recognise the viral DNA as foreign and cut it up in a process called restriction digestion, as it restricts the invading virus from developing. The bacteria's own DNA is immune to this digestion because it is modified for its own protection. Restriction enzymes cut DNA, using specific recognition sequences, leaving overhanging single strands, enabling them to 'stick together'. Hence these became known as 'sticky ends'.

These techniques were so fundamentally important for rDNA technology that Werner Arber and Paul Berg both won Nobel Prizes for their work: Arber received the Nobel Prize in Physiology or Medicine in 1978, together with Hamilton Smith and David Nathan, and Paul Berg shared the 1980 Nobel Prize in Chemistry with Walter Gilbert and Frederick Sanger.

What Paul Berg did that was so innovative was to use a specific restriction enzyme to cut the SV40 DNA into pieces and then use the same restriction enzyme to cut DNA from a bacterial virus, or bacteriophage, (from the Greek *phago*, meaning eating or devouring) called lambda, resulting in the lambda carrying various pieces of SV40 DNA. The final step involved placing the mutant genetic material into a laboratory strain of the *E. coli* bacterium. Because the bacteriophage transfers the inserted DNA into another bacterium, it is called a 'vector', which is the agent used to introduce genes into another organism. This last step, however, was not completed in the original experiment due to the pleas of several fellow investigators who feared the biohazards associated with it.

The SV40 virus is known to cause cancer tumours in mice. Additionally, the *E. coli* bacterium (although not the strain used by

Berg) inhabits the human intestinal tract. For these reasons, the other investigators feared that the final step would create cloned SV40 DNA that might escape into the environment and infect laboratory workers, who might then become cancer victims. These concerns led several leading scientists to send a letter to the President of the National Academy of Sciences (NAS) in which they requested the appointment of an ad hoc committee to study the biosafety implications of this new technology. This committee held a meeting late in 1974 that resolved that scientists should halt experiments involving rDNA until a conference was held to debate these issues.

After this meeting, members sent a letter to *Science*, which became famous as the 'Berg Letter' (Berg *et al*. 1974). In it they wrote:

> There is a serious concern that some of these artificial recombinant DNA molecules could prove biologically hazardous … First, and most important, that until the potential hazards of such recombinant DNA molecules have been better evaluated or until adequate methods are developed for preventing their spread, scientists throughout the world join with the members of this committee in voluntarily deferring the following types of experiments.

They then went on to list these types of experiments, including the introduction of antibiotic resistance or toxin formation, as well as linkage with any potentially oncogenic viruses. They followed this with the caveat:

> The above recommendations are made with the realization that our concern is based on judgments of potential rather than demonstrated risk since there are few available experimental data on the hazards of such DNA molecules.

As requested by the NAS committee, the Asilomar Conference on Recombinant DNA was duly convened in February 1975 with the aim of discussing the potential biohazards and regulation of this technology. About 140 professionals, including biologists, lawyers and physicians, participated and drew up a voluntary set of guidelines to ensure the safety of rDNA. The conference also placed scientific research more into the public domain and saw the first application of the precautionary principle to this technology. This heralded a turning point, both for good and bad, of biotechnology, the effects of which are felt to this day.

The precautionary principle defines actions on issues considered to be uncertain. The principle is used by policy makers to justify discretionary decisions in situations where there is the possibility of harm from making a certain decision when extensive scientific knowledge on the matter is lacking. The principle implies that there is a social responsibility to protect the public from exposure to harm, *when scientific investigation has found a plausible risk.*

During the conference, the principles guiding the recommendations for how to conduct experiments using this technology safely were established. The first principle was that physical and biological containment should be made an essential consideration in the experimental design. However, the second principle, which has been largely overlooked in many cases, was that the effectiveness of the containment should match the estimated risk as closely as possible.

In a paper entitled 'Personal reflections on the origins and emergence of recombinant DNA technology' (Berg and Mertz 2010), Paul Berg reflects on the guidelines implemented as an outcome of the Asilomar conference:

> In the summer of 1976, the National Institutes of Health issued its first set of Guidelines for Research Involving Recombinant DNA. These guidelines and analogous ones from other international jurisdictions along with their updates have been adhered to throughout the world. In the over three decades since adoption of these various regulations for conducting recombinant DNA research, many millions of experiments have been performed without reported incident. No documented hazard to public health has ever been attributable to the applications of recombinant DNA technology. Moreover, the concern that moving DNA among species would breach customary breeding barriers with profound effects on natural evolutionary processes has substantially diminished as research has revealed such exchanges occur in nature as well.

The National Institutes of Health (NIH) guidelines, initially for research funded by the NIH but subsequently extended to cover any research supported by federal funds in the USA, required that different levels of containment be used according to the estimated level of danger. The containment was of two types. The first was biological containment, which depends on the use of weakened strains of the host organisms so that if they were to escape the laboratory they would have difficulty surviving. The second is physical containment, which involves the establishment of four levels of security, including laboratory procedures and, in the case of levels 3 and 4, specialised construction requirements.

Unfortunately, the Asilomar conference and the subsequent NIH guidelines, far from reassuring the public regarding this new technology, resulted in an overwhelming fear of its potential dangers. The media were full of stories of evil-intentioned scientists creating toxin-producing 'superbugs' that could escape from the laboratories in the dead of night, invade the city's drinking water and kill the unsuspecting citizens. In April 1977, *Time* magazine highlighted the issue as its cover story, with a picture showing an evil-looking scientist peering ominously into a test tube containing a pink fluid. This was meant to show that the DNA in the test tube had been mixed with phenol, which is used to extract DNA from bacterial cells. Unfortunately for *Time*, phenol is only pink when it carries impurities that result in the disintegration of DNA. The experiment depicted in this picture would not have worked!

To try to counter these negative attitudes, many of Harvard's senior scientists took to the weekly Saturday marketplace in the town square with models of DNA trying to explain the benefits of rDNA. But the mayor of Cambridge, where the main Harvard campus is located, Alfred Vellucci, banned all work on rDNA in his city, which led to several scientists leaving Harvard to work at universities where such restrictions did not exist.

One of the major concerns expressed by the public, and reflected in the media at the time, was that genetically engineered bacteria could infect humans and cause illness or even death. To counter these fears, some scientists set about determining the fitness of bacteria carrying foreign genes. Many were of the opinion that, as the genomes of bacteria are small, and their genes appeared to be efficiently compact, extra DNA could prove to be a burden and decrease the competitiveness of such bacteria outside the laboratory. As mentioned earlier, if an *E. coli* bacterium expressing a toxin gene from a snake or a scorpion were released into the environment, could it replicate and kill any humans who encountered or swallowed it?

One such scientist was Mark Richmond, Professor of Bacteriology at the University of Bristol, who worked on antibiotic resistance. In an article published with colleagues in *Nature* in 1976 (Hartley *et al.* 1976), he wrote, '*Escherichia coli* K12 is frequently used in genetic engineering experiments. In order to assess the potential hazards of any such experiments it is necessary to study the ability of this organism to survive in the human intestine.' The *E. coli* strain K12 was one of the early strains used in these experiments, because it was a benign version of the *E. coli* found naturally in the human gut. Two laboratory workers, who were engaged in working with *E. coli* K12, carrying plasmids that encoded resistance to the antibiotic, nalidixic acid, became the subjects of their investigations. Plasmids are circular pieces of DNA found in bacteria that are physically separate from and can replicate independently of the chromosomal DNA. They often carry genes for antibiotic resistance and are thus very useful vectors because they can be selected for on media containing that antibiotic.

The two workers took 52 faecal samples over a period of 184 days and looked for the survival of the genetically engineered K12 strains carrying the antibiotic resistance plasmids. None were found and the authors concluded, 'These results suggest that *E. coli* K12 lines are unlikely to become established in the faecal flora of laboratory workers handling these organisms as long as reasonable laboratory procedures are used.' If laboratory workers, handling such bacteria on a daily basis, were not infected, it is highly unlikely that the general public would be at risk.

Another scientist who was concerned about the possibility of *E. coli* becoming a health hazard was Roy Curtiss, Professor of Microbiology at the University of Alabama. In 1978 (Curtiss 1978) he described the use of a derivative of *E. coli* K12, the one tested by Mark Richmond, which was even safer. This strain, *E. coli* EK2 χ1776 contains additional mutations that conferred nutritional requirements, such as diaminopimelic acid, that are not available outside the laboratory. He concluded, 'Thus the introduction of foreign DNA into EK2 host-vector offers no danger.'

In a further analysis of *E. coli* χ1776, Wells *et al.* (1978) found that this strain was so debilitated that it was unable to colonise germ-free rats and mice. They postulated that this strain's inability to colonise conventional animals might have been due to competition from the normal bacterial flora. They therefore introduced it into rats and mice that were maintained at the University of Wisconsin's Gnotobiotic Laboratory under germ-free conditions. The authors found that several attempts to colonise the gastrointestinal tracts of such animals were unsuccessful. 'Viable organisms could not be cultured from either faecal material or tissue samples for up to two months after inoculation.'

In fact, as many scientists working in this field have found, genetically engineered bacteria do not survive well, even under strict laboratory conditions. Hence, such bacteria need to be well cared for and stored at −80°C to ensure their survival. The chances, therefore, of bacteria that express toxin genes killing humans in the environment are remote.

Another reason for scientists' confidence in the safety of rDNA research was the difficulty of introducing plasmid DNA into strains of *E. coli*, a process known as transformation, which is the uptake of DNA by bacteria. Only specific strains were susceptible to transformation and, in a paper published in 1987 (Norgard *et al.* 1978), the authors showed how important various laboratory conditions, such as the use of specific buffering solutions, the presence of magnesium ions and $CaCl_2$ solutions, as well as specific pHs, were for efficient transformation. It was extremely unlikely, therefore, that normal strains of *E. coli*, such as those found in the human gut, could be transformed by rDNA plasmids in their natural habitats.

In addition to the safety of the recipient bacteria, the plasmid vectors that were used to introduce foreign genes into the bacteria were also subject to scrutiny. It was important that these plasmids were not transferable by the natural bacterial 'mating system' known as conjugation. Plasmids that could be transferred by themselves during conjugation were known as self-transmissible. Stanley Falkow, a scientist working on antibiotic-resistant plasmids in the Department of Microbiology at the University of Washington in Seattle, co-published a paper showing the safety of plasmid pBR322 for gene cloning experiments (Bolivar *et al.* 1977). In it they showed that conjugal transfer of this plasmid was significantly reduced relative to its parent plasmid, ColE1. pBR322 and its derivative plasmids have been used safely in bacterial cloning experiments since then.

With all the work described above, which was aimed both at improving the safety of genetic engineering and of showing how useful it could be to both understand how genes worked and to produce useful proteins, the scene was set for commercialisation. Thus, while the work on safety issues was underway, the first biotechnology company, Genentech, was founded in 1976 by Herb Boyer of the University of California, San Francisco, whose group had cloned human somatostatin, and Robert Swanson, a venture capitalist. Boyer had also, with Stanley Cohen, shown how restriction enzymes could cut and ligate genes into plasmid vectors (Cohen *et al.* 1973). In 1978 the company produced synthetic human insulin in bacteria, followed in 1979 with human growth hormone. This venture appeared to gain favour with the media as his smiling, virtuous face graced the front cover of *Time Magazine* on March 9, 1981, only 4 years after the previous cover of an evil counterpart.

The media were reflecting the changed attitude of the public towards genetically engineered bacteria when the product was a medicine that they needed to take in order to be healthy or even to prevent death. Very few patients, when faced with a prescription for life-saving insulin, will ask their doctor how the medicine has been produced. They would also be grateful that they would be taking human insulin, instead of the 'normal', non-genetically engineered pig insulin, which can result in allergic reactions.

This was not the case, however, with genetically engineered bacteria that could affect plants and hence the food we eat. In 1987 Steve Lindow, of the University of California, Berkeley, tested the 'Ice-minus' bacterium he had developed to protect crops such as potatoes and strawberries from frost damage. The bacterium, *Pseudomonas syringae*, when present on plants, can cause ice to form around the ice-plus protein on the bacterial outer cell wall, a process called 'ice nucleation'

(Lindow *et al.* 1989). These strains of *P. syringae* are therefore referred to as 'ice-plus'. The ice damage they cause occurs at temperatures higher than those at which frost would normally occur in the absence of ice-plus bacteria. Ice-minus strains of *P. syrinagae* have had the gene that produces the ice-plus protein deleted and they therefore lack this ice-nucleating protein. If these bacteria are sprayed onto the surface of plants, they can compete with the normal ice-plus strains and decrease the chance of frost damage. These genetically modified *P. syringae* bacteria, known as 'Frostban', were the first GMOs to be field tested and released into the environment. Unfortunately, the heavily protective field gear that Steve and his colleagues wore to show how careful they were being had the opposite effect. In fact, it looked as if the scientists were so afraid of the bacteria that they were protecting themselves from potential damage. Frostban was never marketed.

Another case of using genetically engineered bacteria in the environment also came under public scrutiny, leading to a United States court case. Ananda Chakrabarty, working for General Electric, developed a bacterium, *Pseudomonas putida*, that could break down crude oil. He had developed this strain by stably introducing into it four previously known oil-metabolising plasmids from other strains of *Pseudomonas*. He called his strain 'multi-plasmid hydrocarbon-degrading *Pseudomonas*', but the press dubbed it 'the oil-eating bacteria'. These bacteria became the subject of the first USA patent application for a GMO and, although the patent examiner rejected it on the basis that living things could not be patented, this decision was overturned by the US Court of Customs and Patent Appeals. They stated, 'the fact that micro-organisms are alive is without legal significance for purposes of patent law.'

Throughout the years since those early days of GMOs, there has been little controversy when they are used for the production of pharmaceuticals. The controversy only reared its head when GM crops came onto the market. People believe they are what they eat, and when given a choice of eating food derived from a GM crop or a more 'natural' form, they opt for the latter. Especially as the 'first generation' of such crops favoured the farmer, rather than the consumer. That objection might be true in the developed world, but in developing countries farmers and consumers are often the same person. In addition, the planting of GM crops in such countries often spelt the difference between a reasonable crop and none at all.

As early as 1976, Raymond Valentine's laboratory at the University of California, Davis, was working on the possibility of creating plants that were self-sufficient for nitrogen fixation (Shanmugam and Valentine 1976). This and other examples of GM crops will be dealt with in the next chapter. However, before reading any further, I would be grateful if you would first read the Preface (if you haven't already done so). There I discuss the complex issues of developed versus developing countries and the 'West' versus the 'Rest'.

References

Arber W (1974) DNA restriction and modification. *Progress in Nucleic Acids Research and Molecular Biology* **14**, 1–37. doi:10.1016/S0079-6603(08)60204-4

Berg P, Mertz JE (2010) Personal reflections on the origins and emergence of recombinant DNA technology. *Genetics* **184**, 9–17. doi:10.1534/genetics.109.112144

Berg P, Baltimore D, Boyer H, Cohen HW, Davis RW, Hogness DS, Nathans D, Roblin R, Watson J, Weismann S, Zinder ND (1974) Potential biohazards of recombinant DNA molecules. *Science* **185**, 303–304. doi:10.1126/science.185.4148.303

Bolivar F, Rodriguez RL, Greene PJ, Betlach MC, Heyneker L, Boyer HW, Crosa JH, Falkow S (1977) Construction and characterization of new cloning vehicles. II: a multipurpose cloning system. *Gene* **2**, 95–113. doi:10.1016/0378-1119(77)90000-2

Chang ACY, Nunberg JH, Kaufman RJ, Erlich HA, Schimke RT, Cohen SN (1978) Phenotypic expression in *E. coli* of a DNA sequence coding for mouse dihydrofolate reductase. *Nature* **275**, 617–624. doi:10.1038/275617a0

Cohen SI, Chang A, Boyer H, Helling R (1973) Construction of biologically functional bacterial plasmids in vitro. *Proceedings of the National Academy of Sciences of the United States of America* **70**, 3240–3244. doi:10.1073/pnas.70.11.3240

Curtiss R (1978) Biological containment and cloning vector transmissibility. *The Journal of Infectious Diseases* **137**, 668–675. doi:10.1093/infdis/137.5.668

Curtiss R, Holt RG, Barletta RG, Robeson JP, Saito S (1983) *Escherichia coli* strains producing *Streptococcus mutans* proteins responsible for colonization and virulence. *Annals of the New York Academy of Sciences* **409**, 688–696. doi:10.1111/j.1749-6632.1983.tb26908.x

Goeddel DV, Heyneker HL, Hozumi T, Arentzen R, Itakura K, Yansura DG, Ross MJ, Miozzari G, Crea R, Seeburg PH (1979a) Direct expression in *Escherichia coli* of a DNA sequence coding for human growth hormone. *Nature* **281**, 544–548. doi:10.1038/281544a0

Goeddel DV, Kleid DG, Bolivar R, Heyneker HL, Yansura DG, Crea R, Hirose T, Kraszewski A, Itakura K, Riggs AD (1979b) Expression in *Escherichia coli* of chemically synthesized genes for human insulin. *Proceedings of the National Academy of Sciences of the United States of America* **76**, 106–110. doi:10.1073/pnas.76.1.106

Hartley CL, Petrocheilou V, Richmond MH (1976) Antibiotic resistance in laboratory workers. *Nature* **260**, 558. doi:10.1038/260558a0

Itakura K, Hirose T, Crea R, Riggs AD, Heyneker HL, Bolivar F, Boyer HW (1977) Expression in *E. coli* of a chemically synthesized gene for the human somatostatin. *Science* **198**, 1056–1063. doi:10.1126/science.412251

Lindow SE, Lahue E, Govindarajan AG, Panopoulos NJ, Gles D (1989) Localization of ice nucleation activity and the *iceC* gene product in *Pseudomonas syringae* and *Escherichia coli*. *Molecular Plant-Microbe Interactions* **2**, 262–272. doi:10.1094/MPMI-2-262

Nagata S, Taira H, Hall A, Johnsrud L, Struli M, Ecsödi J, Boll W, Cantell K, Weissman C (1980) Synthesis in *E. coli* of a polypeptide with human leukocyte interferon activity. *Nature* **284**, 316–320. doi:10.1038/284316a0

Norgard MV, Keem K, Monahan JJ (1978) Factors affecting the transformation of *Escherichia coli* strain χ1776 by pBR322 plasmid DNA. *Gene* **4**, 353–354.

Ratzkin B, Lee SG, Schrenk WJ, Roychoudhury R, Chen M, Hamilton TA, Hung PP (1981) Expression in *Escherichia coli* of biologically active enzyme by a DNA sequence coding for the human plasminogen activator urokinase. *Proceedings of the National Academy of Sciences of the United States of America* **78**, 3313–3317. doi:10.1073/pnas.78.6.3313

Shanmugam K, Valentine RC (1976) Solar protein. *California Agriculture* **30**, 4–7.

Shapiro HA, Zwarenstein S (1934) A rapid test for pregnancy on *Xenopus laevis*. *Nature* **133**, 762. doi:10.1038/133762a0

Wells C, Johnson W, Kan C, Kan C, Balish E (1978) Inability of debilitated *Escherichia coli* χ1776 to colonise germ-free rodents. *Nature* **274**, 397–398. doi:10.1038/274397a0

2

GM crops arrive on the scene

As mentioned at the end of Chapter 1, as early as 1976 scientists were starting to think about genetic engineering (GE) of plants. Breakthroughs in recombinant DNA research in the early 1970s led to the production of useful proteins, mainly for medical use. These were based on the introduction into bacteria of novel DNA that coded for beneficial proteins, such as human insulin. In addition, deletions could be made to remove harmful properties of certain bacteria, such as the ice-minus *Pseudomonas*, which could be useful in agriculture and will be discussed in Chapter 3.

An early entrant to this field was Raymond Valentine, who was the scientific founder of the US biotechnology company, Calgene. He, together with his coworkers at the University of California, Davis, was investigating the possibility of making non-leguminous plants fix atmospheric nitrogen. All plants need nitrogen, as it is a major component of amino acids, the 'building blocks' of proteins. Plants obtain nitrogen mainly from the soil, but if it is not present in sufficiently high concentrations, nitrogenous fertilisers have to be used. Not only is this expensive, but excess use can result in run-off into water resources. High levels of nitrogen (as well as phosphorus, also present in many fertilisers) in water can cause algae to grow, in a process called eutrophication. This can lead to the production of toxins, dangerous to human and animal health, as well as 'choking' the water and creating 'dead zones' that prevent fish and other aquatic animals from obtaining oxygen.

There is one group of plants, however, that can fix atmospheric nitrogen into forms that they can readily utilise. These are the legumes, including soybeans, alfalfa, peanuts and clover. Legumes fix nitrogen from the air by attracting and maintaining symbiotic bacteria that carry nitrogen-fixing (*nif*) genes. These genes code for the crucial enzyme nitrogenase, which is responsible for producing the ammonium ion (NH_4^+) from atmospheric nitrogen (N_2). Most of the bacteria capable of fixing nitrogen belong to the genus *Rhizobium* and accumulate NH_4^+ in specially evolved nodules located on the plants' roots. If this ability could be transferred into non-legumes, such as maize, farmers would not have to apply nitrogenous fertilisers to the same extent.

However, in those days there were no known plant vectors capable of transferring genes from bacteria or even another plant into a plant's nucleus. As a result, in a paper published by the journal *California Agriculture*, Valentine and a coworker postulated that plant hybridisation could be used (Shanmugam and Valentine 1976). This is a process whereby breeders cross plants from genetically different parents, usually within the same genus, to produce a hybrid. The hybrid is normally sterile. An example of an interspecific cross is when a male donkey is crossed with a female horse, with the offspring being a sterile mule. Fortunately, agricultural improvements did not have to depend on the processes suggested by Shanmugam and Valentine (1976) because soon after, plant vectors suitable for GE of plants became available.

The search for a plant vector had already begun in the 1970s in many laboratories, but most noteworthy were those in three main competing, and sometimes collaborating, laboratories. In Europe there were Marc van Montagu and Jeff Schell, who shared an office at the University of Ghent in Belgium, while Schell was also a Director in the Max Planck Institute of Plant Breeding, Cologne, Germany. In the USA there was Mary-Dell Chilton, at Washington University in St Louis, where she collaborated with Monsanto who also had a significant team. All three had a background in bacterial genetics.

The basis of their work was the soil bacterium, *Agrobacterium tumefaciens*, which had been shown many decades earlier to cause tumours on the crown section of dicotyledonous plants, where the stem meets the root. These tumours were thus called crown galls. Scientists speculated for many years that the mysterious gall might come from the *A. tumefaciens* bacteria supplying genetic information to the plant to make it start growing a cancerous gall at the site of infection. But how did this take place? Several groups had shown that the presence of a large plasmid was required to induce tumours (Zaenen *et al.* 1974; Larebeke *et al.* 1974) and it became known as the tumour-inducing, or Ti plasmid.

The seminal breakthrough came in 1977 when the Chilton group was (finally) able to show that some DNA from *A. tumefaciens* was in the plants that were producing crown galls (Chilton *et al.* 1977). Their paper, entitled 'Stable incorporation of plasmid DNA into higher plant cells: the molecular basis of crown gall tumorigenesis', showed that the Ti plasmid became integrated into the plant's DNA. But how did this DNA cause tumours?

The answer came when it was found that only a segment of the Ti plasmid became stably integrated into the plant's DNA. They called this segment T-DNA (for transferred DNA) and showed that if foreign genes were inserted into it, they would also be integrated into the plant's DNA (Hernalsteens *et al.* 1980; Matzke and Chilton 1981; Stachel *et al.* 1986). If these genes were expressed, the resulting proteins could change the characteristics of the plant. These plants would therefore have been transformed by the foreign DNA and would be called transgenic plants.

However, two problems still remained. The first was that the T-DNA coded for proteins that caused tumours, the so-called virulence proteins coded for by the *vir* genes. The second was the large size of the Ti plasmid, which made the introduction of foreign genes into it extremely difficult.

Both problems were solved by another competing laboratory, that of Rob Schilperoort from Leiden University (Hoekema *et al.* 1983). The reason that the virulence proteins are a problem is due to the way in which transgenic plants are formed. Scientists grow plant cells in tissue culture and, at an early stage of their development, introduce the foreign genes on plasmids in *A. tumefaciens*. Plants, very conveniently for this purpose, are able to grow from a single cell into a mature plant. This is not possible for animals, including humans. You can't grow a mouse from a single cell. Plants are therefore called totipotent, which means that a single cell is capable of giving rise to any cell type and can result in the formation of a whole plant. The word comes from the Latin *toti*, meaning whole, and the English *potent*. If the *vir* genes were to be transferred into the plant cell, the resultant plant could well be defective. However, the *vir* genes are also required to transfer the T-DNA into plants.

What the Schilperoort laboratory did was to physically separate the T-DNA (which encodes enzymes that produce hormones causing cell division and hence tumour formation) and the *vir* genes (which are needed to transfer the DNA into the plant cell) onto two separate plasmids that were compatible; that is, they could both be present in the same *A. tumefaciens* cell. They put the T-DNA (with the tumour-forming genes removed) into a small plasmid, so that foreign genes could be easily introduced into it. The *vir* genes were put into another, compatible plasmid and both introduced into *A. tumefaciens*. The proteins made by the *vir* genes allow the T-DNA, carrying the foreign genes, to be transferred into the plant cell, but as the *vir* genes themselves are not transferred the plant will not express the harmful tumour-producing proteins. This system, using two plasmids, is known as the binary vector system. However, a transgenic plant, carrying foreign genes, had still not been made by any of the laboratories working on this.

One of the problems was that genes from foreign sources need a promoter, which is a sequence of DNA upstream of the gene, that 'instructs' the enzymes in the cells to transcribe that gene into messenger RNA (mRNA). The mRNA is then translated into a protein. The van Montagu–Schell laboratory published the solution to this when they cloned a plant-specific promoter upstream of the *cat* gene from *Escherichia coli* that confers resistance to the antibiotic, chloramphenicol. The transgenic plant cells, transformed using the *Agrobacterium* system, became resistant to this antibiotic, showing that foreign genes could be expressed in a plant to change its properties (Herrera-Estrella *et al.* 1983).

The Chilton laboratory, then linked with the Bevan and Flavell laboratory in Cambridge UK, published a similar paper using a different plant-specific promoter to achieve resistance to the antibiotic G418, or geneticin, in plant cells (Bevan *et al.* 1983). The van Montagu–

Schell collaboration, the Chilton–Bevan–Flavell team and the Monsanto team announced their successes in GE plants for the first time at the Miami Winter Symposium in January 1983. This was a historic moment in plant science, as the 'race' ended in a photo-finish. It remains a special illustration of how collaboration and competition can lead to efficient progress in science. However, the quest for transgenic plants was far from over. All this work only resulted in transgenic plant cells, which still needed to be regenerated into healthy plants carrying the gene of interest.

The following year the van Montagu–Schell laboratory solved this problem. They published a paper showing the regeneration of transgenic *Nicotiana* (tobacco) plants using, not the binary vector system developed by the Schilperoort laboratory, but a version of the Ti plasmid in which the genes coding for tumour formation had been deleted. This non-oncogenic plasmid, pGV3850, was unable to form a tumour but could express the introduced genes for kanamycin and chloramphenicol resistance in the regenerated plants. Furthermore, these plants were phenotypically and genotypically normal and could set progeny in normal Mendelian fashion with the offspring also expressing the inserted genes (De Block *et al.* 1984).

Soon after this breakthrough, van Montagu and Schell initiated the founding of the first European biotechnology company, Plant Genetic Systems, which was the first to produce an insect-resistant plant in 1987 by GE. That trait will be discussed later in this chapter.

The first GM crop to be approved for commercial release was the Flavr Savr tomato in 1992, produced by the relatively small biotechnology company, Calgene, based in Davis, California. It was approved by the USA Department of Agriculture and went onto the market in 1994. I happened to be visiting Davis, en route to a conference, just when they were launching their new product, which they did with considerable fanfare (it was before the time of the anti-GM crops lobby). They decorated colourful wagons emblazoned with their logo and had every tomato clearly identified with a stick-on label. I naturally bought some and compared the flavour with ordinary tomatoes on sale nearby. I confess I was not 'blown away' by the difference in flavour but they seemed to be somewhat firmer in texture.

What Calgene had done was to slow down the ripening process of the tomato, preventing it from softening, while still allowing the fruit to retain its natural colour and flavour. Mostly, ordinary tomatoes are picked before they have fully ripened and are subsequently artificially ripened using ethylene gas, which acts as a plant hormone. Picking the fruit while unripe allows for easier handling and extended shelf life. In contrast, because they stayed firm for longer, Flavr Savr tomatoes could be ripened on the vine, which was supposed to allow better flavour and colour to develop. Flavr Savr tomatoes produced antisense RNA (RNAi; see Box 2.1; Guo *et al.* 2016), which prevented the production of the enzyme polygalacturonase. This enzyme degrades pectin in the cell wall, resulting in softening of the fruit, which makes them more susceptible to being damaged by fungal infections. Sadly, the tomatoes did not live up to their promise and due to Calgene's relative inexperience in the business of growing and shipping tomatoes, they did not last very long commercially (Charles 2001).

In the UK, the biotechnology company, AstraZeneca, produced a tomato paste using technology similar to the Flavr Savr. Because of the characteristics of the fruit, it was less expensive to produce than conventional tomato paste, resulting in the product being 20% cheaper. Between 1996 and 1999, 1.8 million cans, clearly labelled as GE, were

---

Box 2.1   RNA interference leading to gene silencing

RNAi, or RNA interference, is a process whereby messenger RNA (mRNA) molecules, transcribed from a gene, are prevented from being translated into protein. Small interfering pieces of RNA (siRNA) are made that are complementary, or antisense, to the mRNA. The mechanism probably involves the degradation of double-stranded RNA, when the sense strand is bound by the antisense strand. This process is known as post-transcriptional gene silencing.

   The RNAi pathway is found in many higher organisms and has an important role in nature in defending them from viruses. Gene silencing can be used to increase pest or disease resistance, improve commercial traits and enhance nutritional values (Guo *et al.* 2016).

sold in major supermarkets such as Sainsbury's and Safeway UK. Sadly, due to adverse consumer reaction after an extensive anti-GM campaign, the product was withdrawn (Bruening and Lyons 2000). The reasons for this reaction will be dealt with in Chapter 3.

Another early GM crop that was field tested in China in 1992 was tobacco resistant to tobacco mosaic virus and cucumber mosaic virus (Zhou *et al.* 1995). The trials took place on ~8000 ha in the centre of Henan province. Although the trials showed that tobacco was resistant to the two viruses, the crop was never commercialised. The reasons for this are discussed in Chapter 7.

And then came vitamin A-enriched rice – the so-called 'Golden Rice' – from the Ingo Potrykus laboratory in the Swiss Federal Institute of Technology in Zurich. And what a tale that is, which will be told in further chapters. It started in 2000 when the group reported the engineering of the pro-vitamin A (β-carotene) biosynthetic pathway into carotenoid-free rice endosperm (Ye *et al.* 2000). Not only was this the first GM crop that could be beneficial to consumers instead of farmers, but it involved the cloning of two genes involved in the biosynthetic pathway as opposed to the single gene transfer that had been reported until then. Rice plants possess the machinery to synthesise β-carotene, but parts of it are not expressed in the grains. The Zurich group therefore added a phytoene synthase (*psy*) gene from a daffodil and a phytoene desaturase (*crt*1) gene from a soil bacterium to turn the pathway back on. When the pathway was activated by the addition of the two genes, the physical appearance of the crop changed: the rice grains became a golden-yellow colour due to the presence of β-carotene.

Rice is a major staple food in many parts of Asia, Africa and Latin America. Unfortunately, the rice that is predominantly consumed in these regions lacks pro-vitamin A, which is converted in the body to vitamin A. This has led to a serious public health problem in many countries, as such deficiencies result in blindness, especially in children, and night blindness in pregnant women. It can also diminish the ability of children to fight infections. It would therefore seem obvious that support should be given to Golden Rice, which could be a life-saver in many parts of the developing world. That this has not been the case and that this crop has still not become available to the people who need it most is a saga that will be covered in Chapter 10.

GM crops that have stood the test of time include those that contain insect resistance (IR). These have been widely adopted in agriculture because the introduced trait is effective in making crops resistant to insect attacks, which increases crop yield and decreases the need for chemical insecticides, resulting in both environmental and economic benefits. IR crops were first developed by introducing into the crop species a gene coding for an insect-specific toxin derived from the soil bacterium, *Bacillus thuringiensis*. Hence the crops are often referred to as *Bt* crops. The mode of action of these toxins is described in Box 2.2.

With the increase in scientific activity in the development of GM crops, countries where this was taking place realised that they needed a new set of regulations and guidelines. They would require regulations for

Box 2.2   How *Bt* toxins protect a plant from insects

Bacillus thuringiensis is a soil bacterium that produces a toxin, known as *Bt* toxin, specific for certain insects. *Bt* formulations, consisting of either the bacteria themselves or the protein toxin they produce, have been used safely for many years as non-chemical insecticides. The reason for the insect specificity is that the bacteria produce a form of the toxin that is, in fact, non-toxic but is activated in the gastrointestinal tract of certain insect larvae to form an active toxin. This toxin binds to specific receptors in the lining of the insect's gut, perforating it and causing the rapid death of the larva.

There are many varieties of *B. thuringiensis*, all producing slightly different toxins, which bind to the receptors in different insects. However, many can be grouped together as being toxic to Lepidoptera (butterflies and moths), Diptera (flies) and Coleoptera (beetles), among others. Other animals and humans lack the specific receptors and hence *Bt* is not toxic to them.

Together, the different varieties of *B. thuringiensis* produce hundreds of different toxins that act in slightly different ways, although all bind the receptors and perforate the gut. Thus, if a target insect, such as a maize stem borer, becomes resistant to a given *Bt* toxin, other *Bt* toxins can be used to overcome this resistance. This is analogous to the use by chemical companies of an array of different pesticides when the pest in question becomes resistant to a particular chemical. It is also analogous to medical doctors prescribing a different antibiotic when the bacteria causing a disease become resistant to the first one used.

laboratory experiments, for glasshouse trials, for field trials and for commercialisation. Every country where these were taking place would need to have some form of regulatory authority in place to act in accordance with specific legislature. The basis of such regulations is the premise that GM crops could be harmful to humans, animals or the environment.

Before looking at what these regulations are, let us consider why crops developed by methods other than GE to incorporate novel traits, such as disease and pest resistance, are not similarly regulated? One such technique is called protoplast fusion and is also a type of genetic modification. It occurs when members of two different plant species are fused together to form a new hybrid plant that has the characteristics of both parents. This is done by removing the cell wall of both parents to produce protoplasts and then fusing them together with an electric shock. The formation of the cell wall is induced with hormones and the hybrid cells then grow into plants. This technique has been used to fuse the potato, *Solanum tuberosum*, with its wild relative, *S. brevidens*, to make the former resistant to late blight and the potato leaf roll virus (Helgeson *et al.* 1986).

Another technique arose because one of the biggest obstacles to creating new crop varieties is that conventional plant breeders can only develop new traits from traits that appear naturally in compatible species. What happens when such traits do not exist in a breeding population? In the late 1940s, plant breeders addressed this problem by using one of the newest tools of the day: atomic radiation. Radiation causes random mutations in the plant's DNA. These mutations often lead to plants expressing new traits – some desirable and others not. Plants with desirable traits can then be bred to produce new varieties.

Later, in addition to radiation mutagenesis, chemical mutagens such as ethyl methane sulfonate, methyl methane sulfonate and diethylsulfate were also used to develop new varieties of plants. Hundreds of plants have been developed by these methods, including ruby red grapefruit and varieties of wheat and vegetables. When mutagenesis is used, we have no idea what other changes have been made to the plant's DNA.

None of the above methods are regulated, so crops thus produced can even be sold as organic produce. If countries, such as the EU, choose to regulate GM crops according to the method used, rather than the safety of the end product, surely they should also be regulating crops that have been made by mutagenesis? Even the National Academy of Sciences of the USA has acknowledged that regulating GM crops while giving a pass to products of mutation breeding is not scientifically justified.

However, to return to the regulation of GM crops, different countries use different procedures. In the UK, laws that govern the environment and the use of GMOs are primarily based on EU law. Whether this will continue to be so in a post-Brexit era remains to be seen. The main piece of national legislation that regulates the environment is the *Environmental Protection Act*, which provides the Secretary of State with the power and responsibility to control the deliberate release of GMOs in the UK. The Department for Environment, Food and Rural Affairs is the lead government department for protecting the environment.

In the EU the European Food Safety Authority (EFSA) in independent member states used to make the first decision on whether to allow a GM crop to be released. This had to be approved by the central European authorities in Brussels. Recently, however, the final responsibility for local implementation of GM crops has been handed back to member states, which can now decide whether to opt out from cultivation of a GM crop that was authorised at the EU level. Applications must include an environmental assessment that examines the possible interactions between the GM crop and factors such as soil and other organisms in the ecosystem. To date, no GM crops have been commercialised in the UK. In the EU, only Spain and Portugal grow any GM crops: small amounts of GM maize (ISAAA 2018).

Regulation of GM crops in the USA is divided among three regulatory agencies: the Environmental Protection Agency (EPA), the Food and Drug Administration (FDA), and the US Department of Agriculture (APHIS, USDA). Each of these agencies regulates transgenic crops from a different perspective. The EPA regulates biopesticides, including *Bt* toxins, under the *Federal Insecticide, Fungicide, and Rodenticide Act*. If a crop undergoes GE to carry a gene for a *Bt* toxin, the EPA requires the developer to verify that the toxin is safe for the environment and conduct a food-safety analysis to ensure that the foreign protein is not allergenic. The FDA is responsible for regulating the safety of GM crops that are eaten by humans or animals. According to a policy established in 1992, the FDA considers most GM crops as 'substantially equivalent' to non-GM crops. In such cases, GM crops are designated as 'generally recognised as safe' (GRAS) under the *Federal Food, Drug, and Cosmetic Act* (FFDCA) and do not require pre-market approval. If, however, the insertion of a transgene into a food crop results in the expression of foreign proteins that differ significantly in structure, function or quality from natural plant proteins and are potentially harmful to human health, the FDA reserves the authority to apply more stringent provisions of the FFDCA requiring the mandatory pre-market approval of food additives, whether or not they are the products of biotechnology. The USDA uses the authority of the *Plant Protection Act* to protect agricultural plants and agriculturally important natural resources from damage caused by organisms that pose plant pest or noxious weed risks. The approach of all three agencies is premised on the assumption that regulation should focus on the nature of the products, rather than the process by which they are produced (Pew Trusts 2001).

Most herbicide-tolerant (HT) crops carry a gene that makes them insensitive to the herbicide Roundup®, which contains the active ingredient glyphosate. This chemical inhibits the enzyme EPSPS (5-enol-pyrivylshikimate-3-phosphate synthase) and thus prevents the plant from making essential amino acids. HT crops produce a mutant form of EPSPS that is resistant to glyphosate but still able to make the essential amino acids. When glyphosate is applied to HT crops, the weeds, which contain the normal glyphosate-sensitive form of EPSPS, are killed. One of the advantages of using glyphosate as a herbicide is that it is readily biodegradable by soil microbes, which limits its potential for ground water contamination (Borggaard and Gimsing 2008). The controversy over whether or not glyphosate is carcinogenic will be dealt with in Chapter 6.

Once the HT crops have been harvested, it is possible that trace amounts of the herbicide may remain on the crop. The FDA is responsible for ensuring that any such amounts do not exceed the limits established by the EPA. The latter body evaluates herbicides to ensure that they are safe for human health and the environment when used according to the instructions on the label.

In 1997, the FDA established a voluntary consultation process with GM crop developers to review the determination of 'substantial equivalence' before the crop is marketed. By this they mean that the new food, made from a crop that has undergone GE, can be assessed by comparing it with the food derived from the conventional crop. Thus, the new food will be assessed for toxicity, allergenicity and other pertinent characteristics such as taste, smell etc. If the data provided by the developer are satisfactory, the FDA notifies them that the agency has no further questions regarding the safety assessment. In addition, the FDA reminds the developer that they have a legal obligation to ensure that the food products they bring to market are safe.

Critics raised questions about whether this voluntary consultation process provides adequate assurance that GM crops are safe. In particular, the use of food crops such as corn for the production of non-food products (e.g. pharmaceuticals) does not fall under the FDA's authority. Because of this gap in regulatory authority, the FDA may not perform appropriate oversight until it is too late.

As an answer to this problem, Senator Dick Durbin introduced legislation in 2003 that required any product grown in a food crop to receive pre-market approval, whether or not it was intended to be eaten. Thus, before any pharmaceutical was produced in a food crop, the FDA's Centre for Food Safety and Applied Nutrition would conduct a food-safety analysis to ensure that accidental human exposure to the drug through the food supply will not cause health risks. The bill was passed in 2004. This did not, however, give the FDA new authority to regulate what is known as the Patent Medicine Price Index.

'With the signing of this bill into law, the FDA finally has the tools it needs to ensure that the food on dinner tables and store shelves is safe,' Durbin said. 'The new law will have a dramatic impact on the way the FDA operates – providing it with more resources for inspection, mandatory recall authority, and the technology to trace an outbreak back to its source. I am proud of the work we have done, but our vigilance must continue' (Durbin 2004).

The area of GM crops in the USA is the highest globally, with a total of 75 million hectares planted in 2017, including soybeans, maize, cotton, alfalfa, canola, beet, potato, apples, squash and papaya, which amounts to 40% of the global planting of GM crops (ISAAA 2018). Details of what GM crops are grown where will be discussed in Chapter 3 where we consider the attitudes of Western countries towards GM crops.

However, the cherry on the top of this chapter has to be the awarding of The World Food Prize to Marc van Montagu, Mary-Dell Chilton and Robert Fraley (of Monsanto) in 2013. This prize is often referred to as 'the Nobel Prize for food and agriculture'. According to the Prize's Statement of Achievement, these three scientists were honoured for their 'ground-breaking molecular research on how a plant bacterium could be adapted as a tool to insert genes from another organism into plant cells, which could produce new genetic lines with highly favourable traits' (Global Harvest Initiative 2013).

References

Bevan M, Flavell RB, Chilton M-D (1983) A chimaeric antibiotic resistance gene as a selectable marker for plant cell transformation. *Nature* **304**, 184–187. doi:10.1038/304184a0

Borggaard OK, Gimsing AL (2008) Fate of glyphosate in soil and the possibility of leaching to ground and surface waters: a review. *Pest Management Science* **64**, 441–456. doi:10.1002/ps.1512

Bruening G, Lyons JM (2000) The case of the FLAVR SAVR tomato. *California Agriculture* **54**, 6–7. doi:10.3733/ca.v054n04p6

Charles D (2001) *Lords of the Harvest*. Perseus Publishing, New York.

Chilton M-D, Drummond MH, Merlo DJ, Sciaky D, Montova AL, Gordon MP, Nester EW (1977) Stable incorporation of plasmid DNA into higher plant cells: the molecular basis of crown gall tumorigenesis. *Cell* **11**, 263–271. doi:10.1016/0092-8674(77)90043-5

De Block M, Herrera-Estrella L, Van Montagu M, Schell J, Zambryski P (1984) Expression of foreign genes in regenerated plants and in their progeny. *The EMBO Journal* **3**, 1681–1689. doi:10.1002/j.1460-2075.1984. tb02032.x

Durbin DD (2004) Durbin's Food Safety Bill signed into law. <https://www.durbin.senate.gov/newsroom/press-releases/durbins-food-safety-bill-signed-into-law>

Global Harvest Initiative (2013) The 2013 World Food Prize Laureates: Van Montagu, Chilton, and Fraley. <https://www.globalharvestinitiative.org/2013/09/the-2013-world-food-prize-laureates-van-montagu-chilton-fraley/>

Guo Q, Liu Q, Smith NA, Liang G, Wang MB (2016) RNA silencing in plants: mechanisms, technologies and applications in horticultural crops. *Current Genomics* 17, 476–489. doi:10.2174/1389202917666160520103117

Helgeson JP, Hunt GJ, Haberlach GT, Austin S (1986) Somatic hybrids between *Solanum brevidens* and *Solanum tuberosum*: expression of a late blight resistance gene and potato leaf roll resistance. *Plant Cell Reports* 5, 212–214. doi:10.1007/BF00269122

Hernalsteens J-P, Van Vliet F, De Beuckeleer M, Depicker A, Engler G, Lemmers M, Holsters M, Van Montagu M, Schell J (1980) The *Agrobacterium tumefaciens* Ti plasmid as a host vector system for introducing foreign DNA into plant cells. *Nature* 287, 654–656. doi:10.1038/287654a0

Herrera-Estrella L, Depicker A, Van Montagu M, Schell J (1983) Expression of chimaeric genes transferred into plant cells using a Ti-plasmid-derived vector. *Nature* 303, 209–213. doi:10.1038/303209a0

Hoekema A, Hirsch PR, Hooykaas PJJ, Schilperoort RA (1983) A binary plant vector strategy based on separation of vir- and T-region of the *Agrobacterium tumefaciens* Ti-plasmid. *Nature* 303, 179–180. doi:10.1038/303179a0

ISAAA (2018) 'Global status of commercialized biotech/GM crops 2018: biotech crops continue to help meet the challenges of increased population and climate change'. ISAAA Brief No. 54. The International Service for the Acquisition of Agri-biotech Applications (ISAAA), Ithaca, NY.

Larebeke N, Engler G, Holsters M, van den Elsacker S, Zaenen I, Schilperoort RA, Schell J (1974) Large plasmid in *Agrobacterium tumefaciens* essential for crown gall-inducing ability. *Nature* 252, 169–170. doi:10.1038/252169a0

Matzke AJ, Chilton MD (1981) Site-specific insertion of genes into T-DNA of the *Agrobacterium* tumor-inducing plasmid: an approach to genetic engineering of higher plant cells. *Journal of Molecular and Applied Genetics* 1, 39–49.

Pew Trusts (2001) Guide to U.S. regulation of genetically modified food and agricultural biotechnology products. <https://www.pewtrusts.org/-/media/

legacy/uploadedfiles/wwwpewtrustsorg/reports/food_and_biotechnology/
hhsbiotech0901pdf.pdf>

Shanmugam K, Valentine RC (1976) Solar protein. *California Agriculture* **30**, 4–7.

Stachel SE, Timmerman B, Zambryski P (1986) Generation of single-stranded T-DNA molecules during the initial stages of T-DNA transfer from *Agrobacterium tumefaciens* to plant cells. *Nature* **322**, 706–712. doi:10.1038/ 322706a0

Ye X, Al-Babili S, Klöti A, Zhang J, Lucca P, Beyer P, Potrykus I (2000) Engineering the provitamin A (β-carotene) biosynthetic pathway into (carotenoid-free) rice endosperm. *Science* **14**, 303–305.

Zaenen I, van Larebeke N, Teuchy H, van Montague M, Schell J (1974) Supercoiled circular DNA in crown-gall inducing *Agrobacterium* strains. *Journal of Molecular Biology* **86**, 109–127. doi:10.1016/S0022-2836(74) 80011-2

Zhou RH, Zhang ZC, Wu Q, Fang RX, Mang KQ, Tian YC, Wang GL (1995) Large-scale performance of transgenic tobacco plants resistant to both tobacco mosaic virus and cucumber mosaic virus. In *Proceedings of the Third International Symposium on the Biosafety Results of Field Tests of Genetically Modified Plants and Micro-Organisms*. (Ed. DD Jones) pp. 49–55. University of California, Oakland, CA.

3

The West's stand on GM crops

Where are GM crops grown?

During the period 1992–2016, the EU approved 2404 experimental GM crop field trials for research, while over the same time period the USA approved 18 381 similar trials (ISAAA 2018). In the EU only one non-edible GM crop, insect-resistant maize in Spain and, to a lesser extent, in Portugal, is being grown commercially (FAO 2016), compared with 117 commercial releases in North America and other non-European countries. Of these, 93 have occurred in Canada, which, unlike the EU, whose regulations are based on the method of crop development, focuses on the characteristics of the crop. In Canada, regulators will analyse the new traits, such as disease resistance or drought tolerance, that have been introduced into the crop and determine whether these traits are beneficial or potentially harmful. They will then base their regulatory decision on this assessment and not take into account the methods whereby these traits have been introduced (Smyth and McHughen 2012). A similar scientific approach to regulating the product rather than the process is also adopted in the USA (McHughen and Smyth 2012). It is different in the EU, where the traits are of little consequence but the methods by which the traits have been developed are the most important, if not the overriding, criterion used by regulators to assess whether the crop is approved or not.

These different approaches to regulation are reflected in the distribution of GM crops cultivated globally. Although the global area of GM crops has grown from 1.7 million hectares in 1996 to over 190 million in 2018, the top 10 growers are the USA, Brazil, Argentina, Canada, India, Paraguay, China, Pakistan, South Africa and Uruguay. Australia comes in at number 12 and the only EU country to appear in the list of the top 20 is Spain (ISAAA 2018).

It is fairly easy to understand why Brazil, Argentina and Paraguay have been major adopters of GM technologies, as they grow large amounts of soybeans. Similarly, India, Pakistan and China grow large amounts of cotton. However, why South Africa is the only major, and with Sudan one of only two, countries in Africa growing GM crops, requires further explanation. This will come in Chapter 9.

The six countries that have economically gained the most from GM crops during the first 21 years of commercialisation (from 1996 to 2016) were, in descending order and in billion US$, the USA (80.3), Argentina (23.7), India (21.1), Brazil (19.8), China (19.6) and Canada (8). The other more minor players gained collectively 13.6 to make a total of 186.1 (Brookes and Barfoot 2018).

In 2018, the four major GM crops were soybeans, maize, cotton and canola, with herbicide tolerance (HT) the predominant trait, followed by insect resistance (IR). However, over the years these traits have been stacked, meaning that two or more genes of interest are combined in a single plant. By 2018 the number of stacked HT/IR crops, at 80.5 million hectares, reached almost the same as the 87.5 million hectares grown by HT crops alone, being applied to maize, soybeans and cotton. Some of these crops are even triple stacked, with more than one different *Bt* gene being used to help overcome IR. Canola, not having problems with boring insects, is only HT (ISAAA 2018).

Attitudes towards GM crops

Why did the USA and Europe take such different stances on GM crops? Why are Spain and Portugal the only EU countries growing a GM crop? There is a myth that circulates on both sides of the Atlantic that Americans accepted GM crops without question, while the more precautionary Europeans rejected them. In fact, GMOs went through a period of significant controversy in the USA during the early years starting in the 1980s, as discussed in Chapter 1, even before GM crops arrived on the scene. In addition, approval lengths increased in the 2000s after the StarLink maize recall and the monarch butterfly debate (see later in this chapter).

As mentioned in Chapter 1, both the testing of the ice-minus bacteria with potential to protect plants from frost damage and the granting of patent rights for the development of bacteria that could potentially remove oil spills caused considerable antagonism in the USA.

The science behind the ice-minus bacteria was that there are bacteria of the genus *Pseudomonas* present in the environment that are able to cause water to form ice crystals at temperatures higher than when frost damage would normally occur. Using genetic engineering, scientists removed that gene and, by spraying these ice-minus bacteria onto plants in the field, the GM bacteria would compete with their ice-plus relatives. The result could be a decrease in frost damage. In addition, because such ice-minus strains of *Pseudomonas* can occur naturally by mutation, the laboratory strains would only be mimicking what could be found in nature, except at a higher incidence.

Environmentalists soon started campaigning against the company, Advanced Genetic Systems (Monterey, CA), which developed the 'Frostban' bacteria, claiming that these organisms could cause 'more death and destruction than all the wars ever fought' (Jukes 1987). Jeremy Rifkin, of the Foundation on Economic Trends, began alleging that these bacteria may cause decreased rainfall because they 'could irreversibly affect worldwide climate and precipitation patterns over a long, long period of time' (Skirvin *et al.* 1987). Other objectors claimed that the modified bacteria could cause diseases in humans with impaired immune systems. It was hardly surprising that Frostban was never commercialised.

The granting of the patent to General Electric in 1981 for oil-degrading *Pseudomonas* bacteria caused even more public indignation because many believed that life was sacrosanct and if living organisms could be patented, then where would it end?

Thus, even in the USA, public antagonism to GMOs was simmering in the 1980s. The conflict was further aroused on both sides of the Atlantic when another potential crop arrived on the scene. This was the case of a flounder antifreeze protein gene inserted first into tomatoes (Hightower *et al.* 1991) and then into strawberries (Firsov and Dolgov 1998). As with the case of the General Electric GM bacteria, the fact that again a private sector biotechnology company, DNA Plant Technology Corporation in Oakland, California, was developing these plants, only added to the furore. This was the start of the antagonism towards the private sector for making money out of GM technology.

Of course, introducing one flounder gene into a strawberry does not turn it into a fish, just as transferring the human insulin gene into a bacterium does not turn it into a human. But that didn't deter the opposition and many demonstrators have worn various forms of strawberry–fish hybrid costumes during their protests. In fact, the experiment proved unsuccessful beyond some early positive laboratory results and the idea was soon dropped for all practical applications. However, that didn't stop the wearing of costumes and the display of images that were used with excellent visual effect for years after.

Things quietened down for a while, mainly because the only GMO products on the market were for pharmaceutical uses and, as discussed earlier, when people need medicine, they seldom ask how they are produced. However, this all changed in 1995 when GM crops started to become commercialised in the USA and soon after this anti-GMO attitudes started to re-appear in EU and Scandinavian countries, with terms such as 'frankenfoods' attracting worldwide attention. Part of the negative consumer reaction in the UK was due to the so-called Pusztai affair. This controversy began in 1998 when Arpad Pusztai went public with the initial results of unpublished research he was conducting at the Rowett Institute in Aberdeen, Scotland. He was investigating the possible effects of GM potatoes on rats and claimed that they had stunted growth and their immune systems were suppressed. His comments on a British TV programme caused a storm of controversy and resulted in the Rowett, which had initially supported his work, withdrawing its support. Some of his data were subsequently published (Ewen and Pusztai 1999), but he was criticised by the British Royal Society and others for making an announcement before his experiment was complete or peer-reviewed.

The Rowett Institute subsequently concluded that the data did not support Pusztai's conclusions and a survey by the European Food Safety Authority (EFSA) GMO Panel Working Group on Animal Feeding Trials concluded: 'The studies did not show any biologically relevant differences in the parameters tested between control and test animals' (EFSA GMO Panel Working Group on Animal Feeding Trials 2008). But Pusztai remained a hero in the eyes of many who opposed GMOs and in 2005 he was given a whistle blower award from the Federation of German Scientists (Randerson 2008).

In addition, the development of GM crops in Europe occurred at the same time as initial steps were being taken to integrate national food safety systems into the EFSA. This was politically sensitive because individual countries were losing some of their influence over home-based regulations. I can also argue that the US biotechnology industry blustered its way into the EU, hoping to sell their GM crops to European farmers in this already somewhat hostile regulatory environment. As pointed out by Wesseler and Kalaitzandonakes (2011): 'Never before has a new technology in the field of agriculture been so emotionally debated among stakeholders. In some countries, groups of consumers, politicians and certain non-government organizations (NGOs) have opposed the introduction of GMOs, which they see as a threat to biodiversity, human health, the economy of rural communities, especially in the context of coexistence with organic crops, and as a source of monopolistic power among seed suppliers.'

In his book, *Seeds of Science* (Lynas 2018), Mark Lynas makes some interesting observations regarding the arrival of foods derived from GM crops in Europe in 1996. That year a member of the Green Party in Germany, Bennedict Haerlin, was coordinating an anti-toxics campaign for Greenpeace International when he heard that the first shipment of GM foods, mainly soybeans, was to arrive. He immediately galvanised Greenpeace into action and '… activists swarmed the ships, prevented them temporarily from docking, and unfurled banners calling for a ban on the import of genetically engineered food' (Charles 2001). Friends of the Earth also established a major international campaign against GM food.

Despite this, as Lynas recounts (Lynas 2018), the multinational biotechnology company Monsanto, based in the USA, initially thought it had scored a victory because most people in Europe didn't seem to care. 'Even in Germany, Denmark and the Netherlands, formerly the strong holds of opposition to biotechnology, Greenpeace wasn't able to stir up much public reaction… There was even less reaction in the south in Italy, Spain and France.' But Monsanto's confidence was soon shattered when, rather surprisingly, opposition started rising in the UK. In 1999, the tabloid *The Daily Mirror* published the headline 'Fury as Blair says: I eat Frankenstein food and it's safe'. Under the headline was a photo of the then British Prime Minister Tony Blair with green skin and the caption, 'The Prime Monster' (quoted in Endersby 2012). In fact, as Lynas says, several activists in the UK became the epicentre of the worldwide movement against GMOs. 'By 2000 there were multiple groups campaigning against GM crops: Friends of the Earth, Greenpeace and the Soil Association at the professional end, and Genetix Snowball, Earth First!, GeneWatch and the Genetic Engineering Network at the grassroots'. By 2002 total GM crop field destruction actions carried out by activists were so great that 'there was not much left to destroy'.

Matters weren't helped by the StarLink maize recall. This variety carried the gene for resistance to the herbicide glufosinate and a variant of the *Bt* gene, called *Cry9C*. The Environmental Protection Agency (USA) approved it for use in animal feed but not in human food due to concerns that Cry9C could be allergenic. On 18 September 2000, an organisation called Genetically Engineered Food Alert, whose members included, among others, Friends of the Earth and the Organic Consumers Association, announced that their laboratory had found StarLink maize in Taco Bell products. The products were recalled and a settlement of US$60 million was made to Taco Bell franchisees by the developers, Plant Genetic Systems (Lin *et al.* 2003). This did nothing to endear GM crops to a sceptical public.

Once these anti-GMO sentiments took hold, prejudice against GM crops spread widely. As Albert Einstein said, 'It is harder to crack prejudice than an atom.' One example of a prejudice that is hard to dispel is the view, largely from well-fed Westerners, that agriculture in Africa should be left to smallholder farmers and that commercial farming is 'bad'. There is a rather quaint, but very superior attitude, that 'small is good' and that anything to do with big business, including commercial farming, is not. I have yet to meet a smallholder farmer in Africa who does not aspire to becoming commercial.

Another recurring theme often put forward by Westerners is that poor farmers have to buy GM seed every year and cannot save them for future use. The fact is that many GM crops, such as maize, are hybrids and hybrids have been bought by farmers, commercial and smallholder alike, since they were introduced in the 1930s. The transition from open-pollinated varieties from which famers could save seed, to hybrid, was astonishingly rapid. In Iowa, for instance, the proportion of hybrid maize grew from less than 10% in 1935 to more than 90% in 1939 and by the 1950s the majority of maize throughout the USA was hybrid (Crow 1998). We will return to the question of hybrid crops in Chapter 6.

What were the reasons for this rapid uptake of hybrids, many of which hold true today? First was the increased yield and second the greater uniformity of the crop, which led to easier machine harvesting. But possibly the most important reason at the time was the drought in the USA, known as the 'dust-bowl period', from 1934 to 1936. The hybrid varieties proved to be far more tolerant of these conditions than the conventional ones then in use (Crabb 1947).

Another reason for the antagonism to GM crops in Europe and elsewhere was the general distrust of government agencies due to their handling of mad cow disease and blood samples tainted with human immunodeficiency virus (HIV). Mad cow disease, or bovine spongiform encephalopathy (BSE), is a degenerative neural disease of cows that usually leads to death. It is caused by small proteins called prions becoming misfolded. British politicians assured the public that there was no evidence of any threat to people who ate beef, an assurance that later proved to be incorrect. It turned out that the prions involved in BSE were able to cross the animal–human divide and cause a variant of Creutzfeldt-Jakob disease (Budka and Will 2015).

The HIV-tainted blood scandal occurred when Bayer's Cutter Laboratories discovered that some of their blood products were contaminated with HIV but did not destroy them. As a result, some haemophiliacs in Asia and Latin America tested positive for HIV and developed AIDS (McHenry and Khoshnood 2014). These two issues were among those that diminished the trust that the public might previously have had in government bodies, and even in scientists.

Further antagonism towards GM crops can be seen in articles such as the one written by Ghiselle Karim entitled 'Genetically-modified food: for human need or corporate greed?' (Karim 2013). In it she writes:

'... under capitalism, GMOs are being abused by large agro-corporations, such as Monsanto, to maximise shareholders' profits at the expense of ordinary people around the world.' She goes on to say that the technology does not offer any consumer benefits. Of course, this may well be true for consumers living in industrialised countries, but life is very different in countries, like many in Africa, where the farmer is often also the consumer. Once again, the West versus the Rest.

The World Economic Forum

In 2000 I was invited to speak on GM crops at one of the meetings of the World Economic Forum in Davos. While there I met one of the most prominent anti-GMO activists, Vandana Shiva, and I was asked to participate in a television discussion with her. During the interview, she expounded on the lack of necessity for GM crops as poor people just needed to eat green leafy vegetables to get all the nutrients they needed. 'Tell that to poor people in Africa' I thought to myself, although I didn't comment. However, when she went on to expound on the poor farmers in India who were being driven to suicide after their GM cotton crops failed, I couldn't help myself from saying rather vehemently 'But that is a lie!' The presenter ignored me but we will hear more about Ms Shiva in Chapter 6.

The following year I was invited back to the World Economic Forum, this time to participate in a discussion with Ian Wilmut, one of the scientists who had cloned the sheep 'Dolly', and with one of the scientists leading the human genome project. This discussion was held on the so-called 'free Sunday' and the audience consisted mainly of a small group of heads of state and cabinet ministers.

Ian said they had cloned Dolly simply to see if they could, but that the really important topic in the room was that of GM crops. Sure enough, the audience appeared to agree and I was in the hot seat for most of the session. Questions included issues such as risks to human health and to the environment, and the necessity for GM crops at all.

What the scientists say

These and other questions were being answered at the time by many scientists on many platforms (for overviews see Levidow and Tait 1991;

Kok and Kuiper 2003; König *et al.* 2004). If we are to feed the world's growing population, we will need crops with improved traits not only from an agricultural point of view (drought tolerance, pest and disease resistance etc.) but also from a nutritional one. Classical plant breeding can provide such traits, but only to a limited extent, as it is necessary to obtain fertile offspring with the required characteristics. That process requires breeders to cross plants from the same or closely related species, which is done by transferring the male pollen of one plant to the female organ, the ovule, of another. It can also take a long time, sometimes many years, to achieve the desired results. In addition, a major problem often lies in the fact that the desired characteristic does not exist in any related species. GM technology allows plant breeders to introduce single, or a small number, of genes from any source to confer new properties on the plant.

As mentioned above, one or even a few genes from another organism will not turn that plant into the other organism. In addition, apart from being much quicker than conventional breeding, it is relatively easy to introduce the gene(s) into different varieties of the same crop, thus enhancing biodiversity. For instance, in India there were, as of 2018, 2000 cotton hybrids carrying the *Bt* gene (Vital 2018).

Regarding risks to human health, no food has ever been tested as thoroughly for safety as those derived from GM crops. Before a crop can be released it undergoes a risk assessment that involves determining how the plant was created and what changes were made. Next, a comparative analysis is done to determine whether the plants look the same as the original, whether the composition is similar and whether they have similar agronomical properties and produce similar offspring. Toxicological tests may then be performed, including tests for allergenicity. Indeed, no other food has ever been tested as if it could be a potential toxin. The nutritional value of the GM plant is then determined to ensure it is the same or better than the original. This is done on a case-by-case basis by agencies such as EFSA and the US Food and Drug Agency.

EFSA also requires the toxicological standard 90-day feeding trials of rodents. As many activists claim that 90 days is not long enough, and long-term feeding trials are required, Agnes Ricroch and colleagues from the AgroParisTech and the Institut National de la Recherche Agronomique (INRA) in France undertook an exhaustive review of the 90-day feeding studies (Ricroch *et al.* 2014). They concluded that although the studies were heterogeneous in quality, and as a result some could be considered to be less convincing than others, 'all studies conclude that the commercial GM varieties tested were as safe and nutritious as existing commercial varieties.' They went on to add that as the tests, which have been carried out since 1996, have provided no reasons to doubt the safety of GM crops, 90-day feeding trials are not needed as a routine unless there are specific reasons for doing so. Finally, they stated 'further long-term animal feeding studies or multigenerational feeding studies are not needed either, unless there are specific reasons to do so.' The fact that the EU, unlike the USA, Canada, Australia and others, has imposed mandatory 90-day animal feeding studies indicates that this decision is not based on scientific knowledge, but rather on political issues. The infamous case of the Séralini trials, which purported to show that GM maize causes cancer, will be discussed in Chapter 6.

Regarding environment risk assessment, Katy Johnson and colleagues from the UK published a paper that highlighted the importance of considering not only the potential hazard but the possibilities of exposure to such hazards (Johnson *et al.* 2007). They used the example of the monarch butterfly, which some scientists had shown could be killed by maize pollen landing on milkweed leaves and expressing the *Bt* toxin gene (Losey *et al.* 1999). This caused an enormous public uproar, which only died down some years later when field trials were done that took into account exposure and the environmental behaviour of the butterflies (Sears *et al.* 2001). In fact, in the field trials, it was shown that monarch butterflies avoid pollen-covered milkweed leaves. It was only because the experiments described in the Losey paper were performed in laboratories where the butterflies were fed separately on milkweed leaves with or without pollen that high levels of mortality were recorded. Hence the importance of taking exposure into account. This question of exposure will come up again in the controversy over glyphosate.

A more recent monarch butterfly scare has been attributed to the use of HT maize and the spraying of herbicides that could be killing the milkweeds on which they flourish. Again, scientific facts are required to ascertain whether this could be the case. Boyle *et al.* (2019) have shown that, by studying museum specimen records to chart monarch butterfly and milkweed occurrence over the past century, declines in both began around 1950 and continue to this day. They conclude, 'Whatever factors caused milkweed and Monarch declines prior to the introduction of GM crops may still be at play, and, hence, laying the blame so heavily on GM crops is neither parsimonious nor well supported by data.'

Misinformation or disinformation?

Some years after the World Economic Forum meeting described above, I was invited by Kofi Annan to speak at the United Nations. It soon became apparent that the anti-GM crop lobbyists had reached many of the African countries because I was bombarded with questions such as 'If we eat GM crops will we become sterile?' I later discovered that Greenpeace had a banner that proclaimed this to be true, but it has since been removed from their website.

In addition, the ambassador from Zambia gave a rather convoluted explanation as to why his country, caught in the grip of a terrible drought that was resulting in mass starvation, refused food aid from the USA 'in case' it contained GM maize and 'in case' farmers planted it instead of feeding it to their starving families and thus causing Zambia to lose their 'GM free' status 'in case' they exported their maize to Europe.

This stance is reported to have been promoted in Zambia by the Norwegian Institute of Gene Ecology (GenØk), which was founded in 1998 and is located adjacent to the University of Tromsø. From the outset GenØk was fiercely opposed, both in Norway and internationally, to the use of GM crops, warning that they could have unintended consequences for human health. Apart from organising conferences on biosafey, GenØk staff have travelled around the world promoting perceived risks associated with this technology. In 2003 they organised a course with the inflammatory title 'Regulating a privatized genetic industry which has the potential to destroy the future' (Langberg and Heggdal 2016).

In 2002 they travelled to Zambia when it was in the midst of drought that resulted in a major famine. But it appeared that GenØk was more concerned about food aid coming from the USA that might contain GM maize than of food shortages and the resultant hunger. They alerted Zambian researchers to the potential risks linked to American corn, which led the Zambian government to refuse aid from the USA and hence the argument put forward by the Zambian ambassador to the United Nations (Langberg and Heggdal 2016).

The role of the Precautionary Principle

The Precautionary Principle, as applied to risk-management, states that if an action or policy has a suspected risk of causing harm to the public or to the environment, in the absence of scientific consensus (that the action or policy is not harmful) the burden of proof that it is not harmful falls on those taking an action that may or may not be a risk. Or, to put it more simply, the Precautionary Principle is a modern restatement of the classical Hippocratic oath: 'I will keep them from harm and injustice', which can be rephrased as 'first, do no harm'. Where it differs from this oath is that it should guide the behaviour of institutions and nations and it applies to both human and environmental health (Hanson 2018).

Let us consider this from three points of view: (1) that the Precautionary Principle is essential for the regulation of GM crops, (2) that it is inhibitory to the development of such crops and (3) that, when it is applied properly, GM crops are seen as a useful tool in agriculture.

1. The Precautionary Principle is essential for the regulation of GM crops

Rupert Read's article in *The Ecologist* begins as follows: 'GMOs have been in our diets for about 20 years. Proof that they are safe? No way – it took much, much longer to discover the dangers of cigarettes and trans-fats, dangers that are far more visible than those of GMOs. On the scale of nature and ecology, 20 years is a pitifully short time. To sustain our human future, we have to think long term' (Read 2016).

For these reasons he believes that the Precautionary Principle must be used in decision making regarding GMOs.

Another reason is that, according to him, there is 'no scientific consensus on GMO safety'. He does not, however, give any scientific citations to support this statement. In addition, he considers that risks of ruin 'should be considered far weightier than benefits; because potential benefits of a technology simply cannot outweigh the potential for a truly disastrous outcome, even if the chances of that outcome are relatively small.'

Read's claim that there is no scientific consensus on GMO safety was also stated in a paper published a year earlier by Angelika Hilbeck and colleagues, who included members of GenØK and Vandana Shiva (mentioned in Chapter 6 on the topic of 'dis-communication', being the deliberate distortion of facts). The article, entitled: 'No scientific consensus on GMO safety' (Hilbeck *et al.* 2015) states: 'the claim that there is now a consensus on the safety of GMOs continues to be widely and often uncritically aired…. Claims of consensus on the safety of GMOs are not supported by an objective analysis of the refereed literature.' However, the joint statement mentioned in the article, which has supposedly been signed by over 300 independent researchers (names and affiliations not disclosed), 'does not assert that GMOs are unsafe or safe. Rather the statement concludes that the scarcity and contradictory nature of the scientific evidence published to date prevents conclusive claims of safety, or lack of safety, of GMOs. Claims of consensus on the safety of GMOs are not supported by an objective analysis of the refereed literature.'

This point of view is echoed by Taleb *et al.* (2014) (one of whom is Rupert Read), who argue that GMOs fall under the Precautionary Principle because their risk is systemic, having effects on both the ecosystem and health. They claim that 'GMOs have the propensity to spread uncontrollably', without giving any examples of this happening. With regards to human health, they state that 'foods derived from GMOs are not tested in humans before they are marketed'. They do not explain the ethics of how 'human feeding experiments' would be conducted. And are any new foods such as those derived from mutagenesis or other modern breeding technologies tested on humans?

How has the Precautionary Principle influenced the West's stand on GM crops? We only have to look at the situation in the EU as spelt out in an article published in the Library of Congress (2015): 'The European Union (EU) has in place a comprehensive and strict legal regime on genetically modified organisms (GMOs), food and feed made from GMOs, and food/feed consisting or containing GMOs. The EU's legislation and policy on GMOs, based on the precautionary principle enshrined in EU and international legislation, is designed to prevent any adverse effects on the environment and the health and safety of humans and animals, and it reflects concerns expressed by skeptical consumers, farmers, and environmentalists.'

Thus, at least in the case of the EU, the Precautionary Principle has led to the prevention of the cultivation of GM crops in most of the member countries.

2. The Precautionary Principle is inhibitory to the development of GM crops
'While scientists, being scientists, can rarely find complete unanimity on anything, the majority of scientific reports from international scientific organisations find no evidence that GM crops are more hazardous than their non-GMO counterparts' (Davison and Ammann 2017). Those authors quote the National Academy of Sciences' declaration of no consistent biosafety-related differences between GMOs and non-GMOs. In addition, 'hundreds of studies supported by the European Community revealed no biosafety problems'. They quote the Royal Society of the UK having stated: 'Since the first widespread commercialization of GM produce 18 years ago there has been no evidence of ill effects linked to the consumption of any approved GM crop.' They also refer to many other references in the extensive bibliographies of Weaver and Morris (2005) and Nicolia *et al.* (2013). In the light of these publications, one wonders how authors such as Read (2016) and Hilbeck *et al.* (2015) can make the statement that 'there is no scientific consensus on GMO safety'?

One also wonders, as do Davison and Ammann (2017), whether, as far as GMOs are concerned, the Precautionary Principle is 'simply an excuse by politicians to do nothing and to justify this due to lack of scientific certainty. It provides a justification for halting all progress, while gaining political votes for the protection of the population from all evil (real or imagined).' An earlier article in the same vein stated that in some cases, when decisions regarding GMOs are based on this principle and not on science-based evidence, the responsibility is often passed on to political decision makers (possibly not the most competent to make such decisions) and, eventually, to court rulings (Van Rijssen *et al.* 2015).

Not that the courts can always be relied on to get it right. In the Read (2016) article, the author quotes, in support of his argument for the use of the Precautionary Principle, that in December 2015 the Supreme Court of the Philippines used this principle not only to stop field testing of *Bt* eggplants (also known as brinjals or talong in other parts of the world), but also to temporarily stop any application for field testing, contained use, propagation and importation of GMOs. However, on 26 July 2016, the Supreme Court unanimously reversed its December 2015 ruling. The court declared that the case should have been dismissed in the first place in view of the completion and termination of the *Bt* eggplants field trials and the fact that the biosafety permits had expired in 2012 (ISAAA 2016). It, of course, remains to be seen whether the Precautionary Principle features in future decisions regarding the status of *Bt* eggplants in the Philippines. This will be explored further in Chapter 10.

Regarding the Taleb *et al.* (2014) article, which stated that the Precautionary Principle should be used to prescribe severe limits on GMOs, Ivo Vegter has written a strongly worded rebuttal titled: 'Why Nassim Taleb's anti-GMO position is nonsense' (Vegter 2019). He argues that in most cases the Precautionary Principle considers the dangers an activity could pose but does not consider the potential benefits of that activity. 'We do not forgo driving, even though death is an extreme hazard, since the risk of death is low enough for most of us.' Remember that a hazard is something that has the potential to cause harm, while risk is the chance that this will happen. A major problem with the Precautionary Principle is the difficulty of proving a negative. Thus, establishing evidence of the absence of danger is difficult.

Vegter (2019) uses the analogy of establishing guilt in a court of law, where the prosecution is expected to prove guilt beyond a reasonable doubt before convicting and sentencing someone. As he writes 'the precautionary principle would imprison people upon mere suspicion that they had done something wrong, unless they could prove their innocence beyond a reasonable doubt.'

Taleb *et al.* (2014) used two examples to explain when to use the Precautionary Principle. In the case of nuclear power, they claim that the principle is not warranted. 'Potential nuclear hazards, such as meltdowns and waste, can be large, but are localized and quantifiable. They pose no systemic threat to all of humanity.' GMOs, however, should fall under the Precautionary Principle because of 'the widespread impact on the ecosystem and the widespread impact on health'. In addition, Taleb *et al.* (2014) argue that GMOs have the propensity to spread uncontrollably and therefore their risks cannot be localised, as can nuclear power.

However, as Vegter (2019) points out, GM crops are developed with traits favourable to on-farm cultivation; they do not survive well in the wild. Certainly, there are potential risks with some GM traits, such as the development of resistance by insects or weeds, and these must be mitigated. There is no reason to believe that these could have catastrophic consequences, which is Taleb *et al.*'s threshold for applying the principle.

But is it necessary for a country to abandon the Precautionary Principle as a basis for their support of GM crops as part of their agricultural toolkit? The next section argues against invoking a pro- or anti-Precautionary Principle stance, but instead advocates its proper application.

3.   GM crops can be approved when the Precautionary Principle is applied properly

Let us consider arguments that could allow the Precautionary Principle to validate the use of GM crops.

- Is 'scientific consensus' necessary, bearing in mind that this does not mean 100% unanimity. This is also true for many other issues ranging from climate change to a flat earth. Clearly, as pointed out above, the great majority of scientists, and scientific organisations, agree that GM crops can be grown sustainably.

- Opponents of GM technology inaccurately, and intentionally, exaggerate uncertainty. On the contrary, both the theoretical basis of GM crops and the more than two decades of empirical data derived from experience of growing them indicate that the uncertainty has been reduced to acceptable levels.
- There is a world of difference between the blanket statement of 'risks of GM crops' and specific statements of 'risks of approved GM maize line MON810' or 'risks of *Bt* soybeans in Argentina'. Anti-GMO activists tend to argue in terms of the former generalisation, a risk-focused perspective, while biotechnologists argue from the more specific risk-analysis and data-driven perspective.
- Finally, when benefit–risk ratios are taken into account, based on empirical data, the use of the approved GM crops is, indeed, warranted.

Thus, if the Precautionary Principle is applied based on scientific evidence, and applying the arguments stated above, it would allow GM crops to be approved for cultivation and commercialisation.

Can attitudes be changed?

What then is the current situation in Western countries regarding GM crops? As outlined above, many scientists have published comprehensive data on the safety and economic advantages of GM crops. These data have been critically analysed by many national scientific societies, including the US National Academy of Sciences (National Academies of Sciences, Engineering and Medicine USA 2016), the American Medical Association (Genetic Literacy Project 2012), the UK Royal Society (The Royal Society 2016) and the French Academy of Sciences (Meridian Institute 2002), among others, and all came to the conclusion that GM crops were safe.

However, in May 2019 an interesting article was published by the University of Rochester in the USA entitled: 'Would you eat genetically modified food if you understood the science behind it?' (Knispel 2019). A team of psychologists and biologists from the University of Rochester in the USA, the University of Amsterdam in the Netherlands and Cardiff University in Wales undertook an investigation in the USA, UK and the Netherlands to determine if consumers' negative attitudes to GM crops could be changed if they understood the science better.

They first found that only about one-third of consumers share the views of the national scientific bodies cited above. One reason for this disparity is that activists against GM food have been extremely vocal, using terms such as 'unnatural' or 'frankenfood', which is very different from the rather staid and measured terms used by the scientific community.

They then questioned participants, who responded on a scale of 1 (don't care if foods have been genetically modified), 2 (willing to eat, but prefer unmodified foods) to 3 (will not eat genetically modified foods). In addition, they were asked 11 general science knowledge questions; for instance, whether the universe began with a huge explosion, whether antibiotics kill viruses as well as bacteria, are electrons smaller than atoms and how long it takes for the earth to orbit the sun. The team found that *specific* knowledge about GM foods and procedures was independent of a person's *general* science knowledge, making the former (GM knowledge) nearly twice as useful a predictor of GM attitudes.

As to whether these attitudes could be changed, the answer was 'yes' (McPhetres *et al.* 2019). 'Political orientation and demographics inform attitudes and we can't change those', says McPhetres, the study's lead author. 'But we can teach people about the science behind GMOs, and that seems to be effective in allowing people to make more informed decisions about the products that they use or avoid.' Indeed, they discovered that people's existing knowledge about GM food is the greatest determining factor of their attitudes towards the food. This was more important than all other factors they tested.

Their findings, argues the team, lend direct support for the conclusion that the public's scepticism towards science and technology is largely due to a lack of understanding, or absence of pertinent information. In a comment on this work, McPhetres said that knowledge and appreciation of science is 'the kind of information that people need to make informed decisions about products they use, and the food they eat.' Whether this is true or not remains to be seen.

Before going on to discuss how these attitudes are affecting the uptake of GM crops in the developing world, the next chapter deals with GM crops that are being developed in Africa.

References

Boyle JJ, Dalgleish HJ, Puzey JR (2019) Monarch butterfly and milkweed declines substantially predate the use of genetically modified crops. *Proceedings of the National Academy of Sciences of the United States of America* **116**, 3006–3011. doi:10.1073/pnas.1811437116

Brookes G, Barfoot P (2018) Farm income and production impacts of using GM crop technology 1996–2016. *GM Crops and Food* **9**, 59–89. doi:10.1080/21645698.2018.1464866

Budka H, Will RG (2015) The end of the BSE saga: do we still need surveillance for human prion disease? *Swiss Medical Weekly* **145**, w14212. doi:10.4414/smw.2015.14212

Charles D (2001) *Lords of the Harvest: Biotech, Big Money, and the Future of Food*. Basic Books, Cambridge, MA.

Crabb AR (1947) *The Hybrid-corn Makers: Prophets of Plenty*. Rutgers University Press, New Brunswick, NJ.

Crow FW (1998) 90 years ago: the beginning of hybrid maize. In *Anecdotal, Historical and Critical Commentaries on Genetics*. (Eds JF Crow and WF Dove) pp. 823–928. The Genetics Society of America, Bethesda, MA.

Davison J, Ammann K (2017) New GMO regulations for old: determining a new future for EU crop biotechnology. *GM Crops and Food* **8**, 13–34.

EFSA GMO Panel Working Group on Animal Feeding Trials (2008) Safety and nutritional assessment of GM plants and derived food and feed: the role of animal feeding trials. *Food and Chemical Toxicology* **46**(Suppl. 1), S2–S70. doi:10.1016/j.fct.2008.02.008

Endersby J (2012) *A Guinea Pig's History of Biology*. Chapter 10. Random House, London, UK.

Ewen SW, Pusztai A (1999) Effects of diets containing genetically modified potatoes expressing *Galanthus nivalis* lectin on rat small intestine. *Lancet* **354**, 1353–1354. doi:10.1016/S0140-6736(98)05860-7

FAO (2016) FAOStat. Food and Agricultural Organization. <http://www.fao.org/faostat/en/#data/QC>

Firsov AP, Dolgov SV (1998) Agrobacterial transformation and transfer of the antifreeze protein gene of winter flounder to the strawberry. *Acta Horticulturae* (484), 581–586. doi:10.17660/ActaHortic.1998.484.99

Genetic Literacy Project (2012) American Medical Association reiterates support for GM technology. <https://geneticliteracyproject.org/2012/09/26/american-medical-association-reiterates-support-for-gm-technology/>

Hanson J (2018) Precautionary Principle: current understandings in law and society. In *Encyclopedia of the Anthropocene*. (Eds DA Dellasala and MI Goldstein) pp. 361–366. ScienceDirect (Elsevier), Oxford, UK.

Hightower R, Baden C, Penzees E, Dunsmuir P (1991) Expression of antifreeze proteins in transgenic plants. *Plant Molecular Biology* 17, 1013–1021. doi:10.1007/BF00037141

Hilbeck A, Binimelis R, Defarge N, Steinbrecher R, Szekacs A, Wickson F, Antoniou M, Bereano PL, Clark EA, Hansen M, Novotny E, Heinemann J, Meyer H, Shiva V, Wynne B (2015) No scientific consensus on GMO safety. *Environmental Sciences Europe* 27, 1–6. doi:10.1186/s12302-014-0034-1

ISAAA (2016) 'Global status of commercialized biotech/GM crops: 2016'. The International Service for the Acquisition of Agri-biotech Applications (ISAAA), Ithaca, NY.

ISAAA (2018) 'Global status of commercialized biotech/GM crops 2018: biotech crops continue to help meet the challenges of increased population and climate change'. ISAAA Brief No. 54. The International Service for the Acquisition of Agri-biotech Applications (ISAAA), Ithaca, NY.

Johnson KL, Raybould AF, Hudson MD, Poppy GM (2007) How does scientific risk assessment of GM crops fit within the wider risk analysis? *Trends in Plant Science* 12, 1–5. doi:10.1016/j.tplants.2006.11.004

Jukes T (1987) 'The nonsense about Frostban'. *The Scientist,* 18 May 1987. <https://www.the-scientist.com/opinion-old/the-nonsense-about-frostban-63769>

Karim G (2013) 'Genetically-modified food: for human need or corporate greed?' Defence of Marxism, 1 November 2013. <https://www.marxist.com/gmo-human-need-corporate-greed.htm>

Knispel S (2019) 'Would you eat genetically modified food if you understood the science behind it?' Newscenter, University of Rochester, 24 May 2019. <https://www.rochester.edu/newscenter/genetically-modified-food-consumer-attitudes-science-382922/>

Kok EJ, Kuiper HA (2003) Comparative safety assessment of biotech crops. *Trends in Biotechnology* 21, 439–444. doi:10.1016/j.tibtech.2003.08.003

König A, Cockburn A, Creve RWR, Debruyne E, Grafstroem R, Hammerling U, Kimber I, Knudse I, Kuiper HA, Peijnenburg AACM, Penninks AH,

Poulsen M, Schauzu M, Wal JM (2004) Assessment of the safety of foods derived from genetically modified (GM) crops. *Food and Chemical Toxicology* **42**, 1047–1088. doi:10.1016/j.fct.2004.02.019

Langberg L, Heggdal Ø (2016) 'GenØk: how Norway came to revile GMOs'. Genetic Literacy Project, 18 October 2016. <https://geneticliteracyproject. org/2016/10/18/genok-norway-came-revile-gmos/>

Levidow L, Tait J (1991) The greening of biotechnology: GMOs as environment-friendly products. *Science & Public Policy* **18**, 271–280. doi:10.1093/spp/18.5.271

Library of Congress (2015) Restrictions on genetically modified organisms: European Union. 6th June 2015. <https://www.loc.gov/law/help/ restrictions-on-gmos/eu.php>

Lin W, Price GK, Allen EW (2003) StarLink: impacts on the U.S. corn market and world trade. *Agribusiness* **19**, 473–488. doi:10.1002/agr.10075

Losey JE, Rayor LS, Carter ME (1999) Transgenic pollen harms monarch larvae. *Nature* **399**, 214–216. doi:10.1038/20338

Lynas M (2018) *Seeds of Science: Why We Got It So Wrong on GMOs.* Bloomsbury Sigma, London, UK.

McHenry L, Khoshnood M (2014) Blood money: Bayer's inventory of HIV-contaminated blood products and third world haemophiliacs. *Accountability in Research* **21**, 389–400. doi:10.1080/08989621.2014.882780

McHughen A, Smyth SJ (2012) Regulation of genetically modified crops in USA and Canada: USA overview. In *Regulation of Agricultural Biotechnology: The United States and Canada.* (Eds C Wozniaö and A McHughen) pp. 35–56. Springer Publishers, New York, NY.

McPhetres J, Rutjens BT, Weinstein N, Brisson JA (2019) Modifying attitudes about modified foods: increased knowledge leads to more positive attitudes. *Journal of Environmental Psychology* **64**, 21–29. doi:10.1016/j.jenvp. 2019.04.012

Meridian Institute (2002) French Academy of Sciences announces support for genetically modified crops. <http://www.merid.org/en/Content/News_ Services/Food_Security_and_AgBiotech_News/Articles/2002/12/16/ French_Academy_of_Sciences_Announces_Support_For_Genetically_ Modified_Crops.aspx>

National Academies of Sciences, Engineering and Mediciane USA (2016) Genetically engineered crops: experiences and prospects. <https://www. nap.edu/resource/23395/GE-crops-report-brief.pdf>

Nicolia A, Manzo A, Veronesi F, Rosellini D (2013) An overview of the last 10 years of genetically engineered crop safety research. *Critical Reviews in Biotechnology* **34**, 77–88. doi:10.3109/07388551.2013.823595

Randerson J (2008) 'Arpad Pusztai: Biological divide'. *The Guardian*, 15 January 2008. <https://www.theguardian.com/education/2008/jan/15/academicexperts.highereducationprofile>

Read R (2016) 'The Precautionary Principle: the basis of a post-GMO ethic'. *The Ecologist*, 18 April 2016. <https://theecologist.org/2016/apr/18/precautionary-principle-basis-post-gmo-ethic>

Ricroch AE, Boisron A, Kuntz M (2014) Looking back at safety assessment of GM food/feed: an exhaustive review of 90-day animal feeding studies. *International Journal of Biotechnology* **13**, 230–256. doi:10.1504/IJBT.2014.068940

Sears MK, Hellmich RL, Stanley-Horn DE, Oberhauser KS, Pleasants JM, Mattila HR, Siegfried B, Dively GP (2001) Impact of *Bt* corn pollen on monarch butterfly populations: a risk assessment. *Proceedings of the National Academy of Sciences of the United States of America* **98**, 11937–11942. doi:10.1073/pnas.211329998

Skirvin RM, Kohler E, Steiner H, Ayers D, Laughnan A, Norton MA, Warmund NM (1987) The use of genetically engineered bacteria to control frost on strawberries and potatoes. Whatever happened to all of that research? *Scietia Horticulturae* **84**, 179–189. doi:10.1016/S0304-4238(99)00097-7

Smyth SJ, McHughen A (2012) Regulation of genetically modified crops in USA and Canada: Canadian overview. In *Regulation of Agricultural Biotechnology: The United States and Canada*. (Eds C Wozniaö and A McHughen) pp. 15–34. Springer Publishers, New York, NY.

Taleb NN, Read R, Douady R, Norman J, Bar-Yam Y (2014) The Precautionary Principle (with application to the genetic modification of organisms. arXiv, Cornell University. <https://arxiv.org/abs/1410.5787>

The Royal Society (2016) Is it safe to eat GM crops? <https://royalsociety.org/topics-policy/projects/gm-plants/is-it-safe-to-eat-gm-crops/>

Van Rijssen FW, Eloff JN, Morris EJ (2015) The precautionary principle: making managerial decisions on GMOs difficult. *South African Journal of Science* **111**, 1–9. doi:10.17159/sajs.2015/20130255

Vegter I (2019) 'Why Nassim Taleb's anti-GMO position is nonsense'. *Daily Maverick*, 2 July 2019. <https://www.dailymaverick.co.za/opinionista/2019-07-02-why-nassim-talebs-anti-gmo-position-is-nonsense/>

Vithal BM (2018) 'Bt cotton in India: current scenario'. *Cotton Statistics and News*, 17 July 2018. Cotton Association of India, Mumbai. <https://www.google.com/search?client=firefox-b-d&q=Singh+A+%282018%29+%E2%80%98Cotton+statistics+and+news%E2%80%99.+Cotton+Association+of+India%2C++17+July+17%2C+2018>

Weaver SA, Morris MC (2005) Risks associated with genetic modification: an annotated bibliography of peer reviewed natural science publications. *Journal of Agricultural and Environmental Ethics* **18**, 157–189. doi:org/10.1007/s10806-005-0639-x

Wesseler J, Kalaitzandonakes N (2011) Present and Future EU GMO policy. In *EU Policy for Agriculture, Food and Rural Areas*. 2nd edn. (Eds A Oskam, G Meesters and H Silvis) pp. 403–413. Wageningen Academic Publishersroch, Wageningen, Netherlands.

4

GM crops made in Africa for Africa by Africans

Better bananas

When people in the West think of a banana, they think mainly of one variety, the Cavendish, which is yellow when ripe and is eaten raw. When people in Uganda and other African countries think of a banana they think mainly of plantains or cooking bananas, terms that are used for many varieties of the genus *Musa*, which are usually cooked before eating. In Africa plantains and bananas provide more than 25% of the carbohydrate requirements for over 70 million people (UNCST 2007). But they are subject to several debilitating diseases and pests.

The most devastating disease of bananas and plantains in the Great Lakes region of Africa is banana *Xanthomonas* wilt (BXW), caused by the bacterium *Xanthomonas campestris* pv. *musacearum* (Tripathi *et al.* 2009). The economic effect of the disease is potentially disastrous because it destroys the whole plant, leading to complete yield loss. As there are currently no commercial pesticides, biocontrol agents or resistant cultivars available, the only way that BXW can be managed is by following phytosanitary practices, but the adoption of such practices has been poor because they are labour intensive (Tripathi *et al.* 2009). As a result, Leena Tripathi, then working at the International Institute of Tropical Agriculture (IITA) in Uganda, and now at the same institute in Kenya, and colleagues turned to genetic engineering (GE) for a solution.

They used two genes from the sweet pepper, *Capsicum annuum*, one coding for a hypersensitive response-assisting protein (*Hrap*) and the other for a plant ferrodoxin-like protein (*Pflp*). The hypersensitive response is a mechanism to prevent the spread of infection in plants whereby the infected cell and the invading pathogen are rapidly killed, which prevents the infection from spreading. Ferrodoxins can activate the hypersensitive response when plants are challenged with a bacterial pathogen.

Transgenic bananas were subjected to confined field trials in Uganda by artificially inoculating them with the bacteria. All the non-transgenic plants developed symptoms and eventually wilted completely, but most of the transgenics remained symptomless until harvest. All of the resistant lines produced fruits without symptoms and no viable bacteria were found in them (Tripathi *et al.* 2014).

Tripathi and her colleagues then went on to do an ex-ante economic impact assessment of these bananas in the Great Lakes region. They found that farmers would be willing to adopt the BXW-resistant bananas should they be released and that both consumers and producers would benefit. In fact, the consumers would benefit twice as much as the farmers due to the price reduction from the excess, stable and continuous supply of bananas (Ainembabazi *et al.* 2015). Sadly, as will be discussed in Chapter 9, these plants have not yet become available to farmers in the region, whose bananas, as a result, continue to be afflicted by this potentially curable disease.

Another problem facing banana farmers in the region is plant-parasitic nematode infestation. The nematodes can be controlled by application of pesticides, but these are environmentally damaging. Although conventionally bred nematode-resistant bananas have been obtained, these are only resistant to one of the many different plant-parasitic species of nematodes attacking bananas. Tripathi and her colleagues once again came up with a solution. They decided to disrupt the feeding of the nematodes by introducing into the bananas the gene coding for a cystatin that inhibits the cysteine proteinases, major digestive enzymes in nematodes (Roderick *et al.* 2012). A second approach was to reduce nematode invasion and hence the damage they cause. This can be done by using non-lethal synthetic peptides that have this inhibitory effect (Liu *et al.* 2005). In field trials of transgenic bananas expressing genes for both these proteins, up to 99% nematode resistance was found (Tripathi *et al.* 2015). Yield was up to 186% higher compared with the non-transgenic plants due to less root damage. Again, farmers in the region will have to wait for politicians to catch up with the needs of their people.

Finally, Tripathi's team has used the relatively new gene editing approaches (these will be discussed in Chapter 11) to develop bananas resistant to the endogenous banana streak virus (Tripathi *et al.* 2019).

Another team working on bananas is headed by James Dale from the Queensland University of Technology in Brisbane, Australia, working very closely with scientists in Uganda. Their aim is to produce a 'vitamin A super banana', with generous funding from the Bill and Melinda Gates Foundation. Vitamin A deficiency remains one of the world's major public health problems despite food fortification and supplements strategies. This team have developed biofortified bananas with enhanced levels of pro-vitamin A (PVA) using GE. Initial experiments were done using the Cavendish variety, but cultivars suitable for Ugandan diets will be used in Africa (Waltz 2014; Paul *et al.* 2017).

Initially the team used transgenes (i.e. genes derived from species other than the *Musa* genus), but more recently they are using cis-genes derived from bananas themselves. Because bananas are triploids, with three copies of all chromosomes, there is a high level of male and female sterility. From an environmental perspective, this is an advantage, as gene flow into wild *Musa* species is highly improbable. Therefore, say the authors, 'genetic changes in bananas may be compatible with organic farming' (Dale *et al.* 2017).

Many of the principles of organic farming overlap with the primary principles of the team's banana project. Their aim is to eliminate pesticides and allow the sustainable production of landraces, although for the biotechnologists this is done using genetic improvement. In Africa the majority of bananas (up to 85%) are produced by smallholder, subsistence farmers and consumed locally. Most are grown without fertilisers, pesticides or herbicides. They can therefore be described as 'organic'. Of course, under current definitions, genetically engineered improved cultivars would not be considered organic, but what could be more organic than a piece of DNA? Perhaps it is time for the definition of 'organic' to be modified so that poor farmers and consumers in countries such as Africa can benefit from these improved varieties (Dale *et al.* 2017).

Insect-resistant cowpeas

Another crop, known to the Western world as black-eyed peas, is cowpea (*Vigna unguiculta* Walp.), the most important legume food crop grown in subSaharan Africa. Cowpea serves as a major source of dietary protein, being consumed as a fresh leafy vegetable, soft pods as well as grain. In West Africa it is also the main forage crop (Timko and Singh 2008). It is a legume and therefore its roots host rhizobial bacteria that fix atmospheric nitrogen that can be used by crops either planted later in the same soil or by intercropping. It is well adapted to sandy soil, drought-tolerant and therefore suited to low rainfall regions. Hence it is an important crop in semiarid regions and is considered to be the most important food grain legume in tropical Africa where it is eaten by some 200 million people. It is also used as fodder for livestock and provides cash income (Timko and Singh 2008).

Unfortunately, the production of cowpeas is constrained by several insect pests, with the *Maruca vitrata* pod borer (MPB) being one of the most devastating. As there are no genes for resistance to MPB in the cowpea germplasm, traditional plant breeding cannot be used to control this pest (Fatokun 2002). Consequently, GE is the best solution. Several scientists, together with the African Agricultural Technology Foundation (AATF), have been working on the development of cowpea resistance to MPB for several years. AATF obtained the *Bt Cry1Ab* gene from Monsanto and the transformation of cowpea, variety IT86D-1010, was carried out by Commonwealth Scientific and Industrial Research Organisation (CSIRO) in Australia under the leadership of TJ Higgins. Cowpea events, being varieties of a GM crop in which a specific transgene has been introduced in a specific place in the genome, were sent to Africa (Nigeria, Burkina Faso, Ghana and Malawi) for efficacy testing. The confined efficacy tests were conducted under severe artificial infestations. One event was identified as being potentially useful because it had only one copy of the *Cry1Ab* gene. It showed nearly complete resistance to MPB, increased the number of pods per plant by 1.6–13-fold and grain yield by several fold (I. Abdourhamane, pers. comm., August 2019).

Event 709A was used to backcross the *Cry1aB* gene into farmers' preferred varieties, and, depending on the pressure of *Maruca*, the new *Bt*-cowpea lines out-yielded conventional cowpeas by from 20% to more than 100%. In order to prevent resistance build-up to a single gene, a second *Bt*, *Cry2Ab*, has been used to transform cowpea and the best six events are undergoing efficacy tests in West Africa. The genes *Cry2Ab* and *Cry1Ab* will be stacked so that both are present in a plant, in order to prevent the build-up of resistance in the MPB (I. Abdourhamane, pers. comm., August 2019).

TJ Higgins and his teams at the CSIRO in Canberra and in Africa at the AATF, as well as both the Samaru Institute of Agricultural Research, and the IITA in Nigeria, are using another approach to prevent insect resistance to *Bt* toxins. In addition to using a variety of *Bt Cry* genes, they are using *Bacillus thruringiensis* genes encoding vegetative insecticidal proteins (Vips). However, genes from bacteria are often poorly expressed in higher organisms such as plants because the former are rich in adenine and thymine bases in their DNA and often produce fairly unstable messenger RNA (mRNA). As a result, *Cry* and *vip3Ba* genes were modified to improve their expression (Perlak *et al.* 1991; Das *et al.* 2016).

When they tested their reconstructed *vip3Ba* gene in transgenic cowpeas, the team showed that the Vip3Ba protein levels were high enough to cause *Maruca* larvae death and protect the plants completely (Bett *et al.* 2017). In their paper they state: 'Thus, it is proposed that the *vip3Ba* gene is an attractive candidate to complement *cry* genes in the development of *Bt* cowpeas resistant to MPB, thereby contributing to an effective strategy for insect resistant management. This combination would considerably reduce the likelihood of development of resistance in MPB, help to increase yield and income, and reduce the dependency on pesticides.'

At the end of January 2019, the Nigerian National Biosafety Management Agency announced the approval for commercial production of *Bt* pod borer resistant cowpea. The environmental release of the *Bt* pod borer resistant cowpea allowed these varieties to be included in the National Variety Performance trials and the production of foundation seeds on a large scale. Finally, on 12 December 2019 the government approved the commercial release of this crop – an extremely important step forward both for the country and for Africa (ISAAA 2019). This will be discussed further in Chapter 9.

Virus-resistant crops

Africa is home to some plant viruses not found in many other countries. Some years ago, scientists at the Donald Danforth Plant Science Centre in St Louis, Missouri, teamed up with scientists in Uganda and Kenya to try to solve the problem of viruses attacking cassava (*Manihot esculenta*). Cassava is a very important food crop throughout East Africa (Omamo *et al.* 2006), but cassava brown streak disease (CBSD) has become a major constraint to its production. It is caused by two viruses, *cassava brown streak virus* (CBSV) and *Ugandan cassava brown streak virus* (UCBSV), transmitted by whiteflies (*Bemisia tabaci*). The disease can reach epidemic proportions, with incidences reported to be as high as 93% of farmers' fields in a survey carried out in western Kenya (Mware *et al.* 2009). The roots are the edible parts of cassava and CBSD causes necrosis of the roots, rendering them practically inedible. Some success in developing virus tolerance has been achieved by conventional breeding, but no effective resistance has been found. The scientists involved have therefore turned to RNAi (see Chapter 2, Box 2.1) technology to achieve resistance using GE to silence a specific gene.

A genetic construct, in which more than one gene is introduced in the same position in the genome, aimed at conferring resistance to both CBSV and UCBSV was expressed in cassava and found to confer resistance to both viruses under field conditions with high disease pressure (Yadav *et al.* 2011). If Uganda succeeds in passing biosafety legislation in its parliament, and now that Kenya has lifted its ban on GM crops, the potential benefits of GM cassava to smallholder farmers will be substantial (Tomlinson *et al.* 2018). It has been estimated that the value could be as high as US$436 million for western Kenya and US$790 million for Uganda over a 35-year period beginning in 2025 (Taylor *et al.* 2016).

Another damaging virus that attacks maize only in Africa and its neighbouring islands, is maize streak virus (MSV) transmitted by a leafhopper, *Cicadulina mbila*. Ed Rybicki, South Africa's leading plant virologist and I joined our teams to develop GM maize resistant to MSV. We succeeded using several strategies based on the fact that the virus needs host proteins to replicate itself. By mutating these, while still leaving them active in maize DNA replication, we were able to generate lines of maize that were almost immune to MSV (Shepherd *et al.* 2007). The paper caught the attention of *Science*, which published an article with the sub-heading: 'The first genetically modified crop developed entirely in Africa is gearing up for field trials. Its success would be a milestone' (Sinha 2007). Sadly, those field trials have never taken place. Our funder at the time was the South African seed company, Pannar Seed, which simply could not afford the expense involved. Some years later they were taken over by the multinational seed company, Pioneer, but they too were unable to undertake such trials. Their reason was that MSV-resistant maize would only benefit African farmers who were too poor to afford to pay for the more expensive GM seeds. The extra payment would have to be paid to Pioneer in order for them to recoup the costs of field trials. The seeds remain in the freezers at the University of Cape Town and Pannar Seed.

Drought-tolerant maize

Both northern and southern Africa are projected to experience increased drying during the 21st century and could become more vulnerable to the effects of drought. The daily maximum temperature is projected to increase by up to 8°C, precipitation could decrease while warm spells increase (Gan *et al.* 2016). With this in mind, several groups are investigating ways to improve the drought tolerance of crops. One such initiative is the Water Efficient Maize for Africa (WEMA) project under the management of the AATF. Monsanto donated, royalty free, the gene *cspB*, which encodes a cold-shock protein from the bacterium *Bacillus subtilis*. This protein has been shown to confer tolerance to dehydration in transgenic maize (Chang *et al.* 2014). The gene has been introduced into African maize varieties by the International Maize and Wheat Breeding Organization (CIMMYT), which, although based in Mexico, has centres in Africa.

Through the WEMA Project (now known as the TELA [TELA is derived from the Latin word *tutela*, which means 'protection'] Maize Project), three transgenic traits encoded by the drought-tolerant transgene (*cspB*, DroughtGard˙, MON87460), the cold-shock protein gene (*CspB*) and three *Bt* genes (MON810 (*Cry1Ab*) and MON89034 (*Cry1A.105* and *Cry2Ab2*)) were accessed royalty free for smallholder farmers in Africa by the AATF. The *Bt* genes have been stacked with the DroughtGard˙ gene to provide several variety options for farmers who operate in drought-prone environments.

Confined field trials (CFTs) were carried out in Kenya and Uganda for five seasons to test the efficacy of the *Bt* MON810 gene in controlling the spotted stemborer (*Chilo partellus*) and the African stemborer (*Busseola fusca*). Under conditions of artificial infestation, the maize hybrids containing the *Bt* gene yielded, on average, 52% more than the isogenic hybrids without the gene (Kyetere *et al.* 2019). Similarly, CFTs carried out with the stacked drought-tolerant and insect-protection (DroughtGard + *Bt* MON810) traits under natural infestation of fall armyworm (FAW: *Spodoptera frugiperda*) and stemborers in Ethiopia, Mozambique and Uganda, and under natural infestation of FAW and artificial infestation with stemborer larvae in Kenya and Tanzania showed that the *Bt* MON810 trait gave partial but significant control against FAW and full control of stemborers. For example, preliminary results of the trials carried out in Mozambique and Uganda under natural infestation of both FAW and stemborers showed that all the transgenic maize hybrids except one realised 9–98% yield advantage over the non-transgenic isogenic hybrids based on the level of infestation.

The FAW is a new insect pest, native to the Americas but recently reported in Africa, where it is ravaging staple crops, particularly maize, causing serious crop destruction with estimated maize yield losses of 8.3–20.6 million tons, worth US\$2.48–6.19 billion in 12 African countries (CABI 2017). It was first reported in Nigeria in 2016, but has since spread to over 40 countries in Africa, thus posing a major threat to food and nutrition security in Africa.

In South Africa, which has a long history of commercialising biotech crops, five TELA˙ hybrids have been commercialised to smallholder farmers since 2016. Farmers are currently growing these with good protection against stemborer and FAW.

In addition to this work, my laboratory has also been involved in developing potentially drought-tolerant maize. We have used three genes, driven by a stress-inducible promoter, from the resurrection plants, *Xerophyta viscosa* and *X. schlecteri*. Resurrection plants are so named because they can withstand the loss of up to 95% of their relative water content and 'resurrect' to normality within ~72 h of watering (Mundree *et al.* 2000) The three genes code for a stress-associated protein, an aldose reductase and a peroxiredoxin (Mundree *et al.* 2000; Garwe *et al.* 2003; Govender *et al.* 2016). In 2018 we were awarded a grant by the Technology Innovation Agency of the South African Department of Science and Innovation to take the project to glasshouse trials and, if successful, to field trials. We have partnered with Klein Karoo Seed Marketing (K2), situated in the small town of Oudtshoorn, some 400 km from Cape Town. They have highly skilled seed scientists with experience in growing GM crops, including maize, in their biosafety compliant glasshouses and in the field. Time will tell if these plants do, indeed, tolerate drought conditions.

There can be no question about the importance of drought-tolerant crops for both Africa and the rest of the world that will be suffering from the effects of climate change over the next few years, and for decades to come, if we don't act decisively now. If these potentially drought-tolerant crops, developed through GE, are not allowed to reach farmers because of anti-GMO sentiments emanating from Europe (see Chapters 9 and 10) it will be nothing less than 'a crime against humanity' (an expression first used by Patrick Moore, past-President of Greenpeace Canada).

Late blight-resistant potato

Late blight of potatoes (*Solanum tuberosum*), caused by *Phytophthera infestans*, is a devastating disease of potato, with ~15–30% loss of annual yield in subSaharan Africa, affecting mainly smallholder farmers (Hareau *et al.* 2014). Three late blight resistance genes have been transferred from wild relatives, *S. bulbocastanum* and *S. venturi*, into farmer-preferred potato varieties 'Desiree' and 'Victoria'. In field tests in Uganda, transgenic events grew normally and showed pathogen resistance, compared with their non-transgenic equivalents. Two of the transgenics gave yields of 29 and 45 tonnes/ha, respectively, which is an almost 3–4-fold increase over the national average (Ghislain *et al.* 2019). As the late blight pathogen has very little genetic diversity in subSaharan Africa, with the 2_A1 lineage being predominant (Njoroge *et al.* 2016), it is likely that resistance will be long-lasting, which could bring significant income to smallholder farmers in the region. The introduction of these genes from wild species through conventional breeding would take decades of crossing and selection due to the genetic drag of other possibly negative genes from the wild species. Such negative effects would be difficult to eliminate, because cultivated potatoes are tetraploid (Ghislain *et al.* 2019).

It is clear from the work described in this chapter that African scientists are developing many extremely useful GM crops. Sadly, most have not seen the light of day as far as farmers and consumers are concerned due to lack of government approval for the commercialisation of GM crops. In the next chapter we will see what economists and political scientists have to say about this.

References

Ainembabazi JH, Tripathi L, Rusike J, Adboulaye T, Manyong V (2015) Ex-ante economic impact assessment of genetically modified banana resistant to Xanthomonas wilt in the Great Lakes Region of Africa. *PLoS One* **10**, e0138998. doi:10.1371/journal.pone.0138998

Bett B, Gollasch S, Moore A, James W, Armstrong J, Walsh T, Harding R, Higgins TJV (2017) Transgenic cowpeas (Vigna unguiculata L. Walp) expressing Bacillus thruingiensis Vip3Ba protein are protected against the Maruca pod borer (Maruca vitrata). *Plant Cell, Tissue and Organ Culture* **131**, 335–345. doi:10.1007/s11240-017-1287-3

CABI (Centre for Agriculture and Bioscience International) (2017) Fall armyworm: impacts and implications for Africa. <www.invasive-species. org/fawevidencenote>

Chang J, Clay DE, Hansen SA, Clay SA, Schumacher TE (2014) Water stress impacts on transgenic drought-tolerant corn in the northern great plains. *Agronomy Journal* **106**, 125–130. doi:10.2134/agronj2013.0076

Dale J, Paul J-Y, Dugdale B, Harding R (2017) Modifying bananas: from transgenics to organics? *Sustainability* **9**, 333. doi:10.3390/su9030333

Das A, Datta S, Sujayanand GK, Kumar M, Singh AK, Arpan SA, Ansari J, Kumar M, Faruqui L, Thakur S, Kumar PA, Singh NP (2016) Expression of chimeric Bt gene, Cry1Aabc in transgenic pigeonpea (cv. Asha) confers resistance to gram pod borer (Helicoverpa armigera Hubner). *Plant Cell, Tissue and Organ Culture* **127**, 705–715. doi:10.1007/s11240-016-1131-1

Fatokun CA (2002) Breeding cowpea for resistance to insect pests: attempted crosses between cowpea and *Vigna vexillata*. In *Challenges and Opportunities for Enhancing Sustainable Cowpea Production*. Proceedings of the World Cowpea Conference III held 4–8 May 2000 at the International Institute of Tropical Agriculture (IITA), Ibadan, Nigeria. (Eds CA Fatokun, SA Tarawali, BB Singh, PM Kormawa and M Tamòs) pp. 52–61. IITA, Ibadan, Nigeria.

Gan TY, Ito M, Hülsmann S, Qin X, Lu XX, Liong SY, Rutschman P, Disse M, Koivusalo H (2016) Possible climate change/variability and human impacts, vulnerability of drought-prone regions, water resources and capacity building in Africa. *Hydrological Sciences Journal* **61**, 1209–1226. doi:10.1080/02626667.2015.1057143

Garwe D, Thomson JA, Mundree SG (2003) Molecular characterization of XVSAP1, a stress responsive gene from the resurrection plant Xerophyta viscosa Baker. *Journal of Experimental Botany* **54**, 191–201. doi:10.1093/jxb/erg013

Ghislain M, Byarugaba AA, Magembe E, Njoroge A, Rivera C, Román ML, Tovar JC, Gamboa S, Forbes GA, Kreuze JF, Barekye A, Kiggundu A (2019) Stacking three late blight resistance genes from wild species directly into African highland potato varieties confers complete field resistance to local blight races. *Plant Biotechnology Journal* **17**, 1119–1129. doi:10.1111/pbi.13042

Govender K, Thomson JA, Mundree S, Eisayed AI, Rafudeen MS (2016) Molecular and biochemical characterisation of a novel type I peroxiredoxin (XvPrx2) from the resurrection plant Xerophyta viscosa. *Functional Plant Biology* **43**, 669–683. doi:10.1071/FP15291

Hareau G, Kleinwechter U, Pradel W, Suarez V, Okello J, Vikraman S (2014) 'Strategic assessment of research priorities for potato: Lima (Peru)'. CGIAR Research Program on Roots, Tubers and Bananas (RTB) Working Paper 2014–8. <https://cgspace.cgiar.org/handle/10568/69246?show=full>

ISAAA (2019) 'Nigeria commercializes pod borer resistant cowpea, its first GM food crop'. ISAA Crop Biotech Update, 18 December 2019. <http://www.isaaa.org/kc/cropbiotechupdate/article/default.asp?ID=17899>

Kyetere D, Okogbenin E, Okeno O, Sanni K, Munyaradzi J, Nangayo F, Kouko E, Oikeh S, Abdourhame I (2019) The role and contribution of plant breeding and plant biotechnology to sustainable agriculture in Africa. *Afrika Focus* **32**, 83–109.

Liu B, Hibbard JK, Urwin PE, Atkinson HJ (2005) The production of synthetic chemodisruptive peptides in planta disrupts the establishment of cyst nematodes. *Plant Biotechnology Journal* **3**, 487–496. doi:10.1111/j.1467-7652.2005.00139.x

Mundree SG, Whittaker A, Thomson JA, Farrant JM (2000) Cloning and characterization of an aldose reductase homologue from the resurrection plant Xerophyta viscosa Baker by functional sufficiency in Escherichia coli. *Planta* **211**, 693–700. doi:10.1007/s004250000331

Mware B, Narla R, Amata R, Olubayo F, Songa J, Kyamanyua S, Ateka EM (2009) Efficiency of cassava brown streak virus transmission by two whitefly species in coastal Kenya. *Journal of General and Molecular Virology* **1**, 40–45.

Njoroge AW, Tusiime G, Forbes GA, Yuen JE (2016) Displacement of US-1 clonal lineage by a new lineage of Phytophthora infestans on potato in Kenya and Uganda. *Plant Pathology* **65**, 587–592. doi:10.1111/ppa.12451

Omamo SW, Diao X, Wood S, Chamberlin J, You L, Benin S, Wood-Sichra U, Tatwangire A (2006) 'Strategic priorities for agricultural development in East and Central Africa'. Research Report 150. International Food Policy Research Institute (IFPRI), Washington, DC.

Paul J-Y, Khannal H, Kleidon J, Hoang P, Geijskes J, Daniells J, Zaplin E, Rosenberg Y, James A, Mlalazi B, Deo P, Arinaitwe G, Namanya P, Becker D, Tindamanyire J, Tushemereirwe W, Harding R, Dale J (2017) Golden bananas in the field: elevated fruit pro-vitamin A from the expression of a single banana transgene. *Plant Biotechnology Journal* **15**, 520–532. doi:10.1111/pbi.12650

Perlak FJ, Fuchs RL, Dean DA, McPherson SL, Fischhoff DA (1991) Modification of the coding sequence enhances plant expression of insect control protein genes. *Proceedings of the National Academy of Sciences of the United States of America* **88**, 3324–3328. doi:10.1073/pnas.88.8.3324

Roderick H, Tripathi L, Babirye A, Wang D, Tripathi JN, Urwin PE, Atkinson HJ (2012) Generation of transgenic plantain (Musa spp.) with resistance to plant pathogenic nematodes. *Molecular Plant Pathology* **13**, 842–851. doi:10.1111/j.1364-3703.2012.00792.x

Shepherd DN, Mangwende T, Martin DP, Bezuidenhout M, Kloppers FJ, Carolissen CH, Monjane AL, Rybicki EP, Thomson JA (2007) Maize streak virus-resistant transgenic maize: a first for Africa. *Plant Biotechnology Journal* **5**, 759–767. doi:10.1111/j.1467-7652.2007.00279.x

Sinha G (2007) GM technology develops in the developing world. *Science* **315**, 182–183. doi:10.1126/science.315.5809.182

Taylor NJ, Sekabira HA, Sibiko KW, Gua A, Lynam JK (2016) *Disease-resistant GM cassava in Uganda and Kenya during a pandemic.* Biosciences for Farming in Africa. Banson, Cambridge, UK.

Timko MP, Singh B (2008) Cowpea, a multifunctional legume. In *Genomics of Tropical Crop Plants: Plant Genetics and Genomics: Crops and Models,* Vol. 1. (Eds PH Moore and R Ming) pp. 95–114. Springer, New York, NY.

Tomlinson KR, Bailey AM, Alicai T, Seal S, Foster GD (2018) Cassava brown streak disease: historical timeline, current knowledge and future prospects. *Molecular Plant Pathology* **19**, 1282–1294. doi:10.1111/mpp.12613

Tripathi L, Mwangi M, Abele S, Aritua V, Tushemereirwe WK, Bandyopadhyay R (2009) Xanthomonas wilt: a threat to banana production in east and central Africa. *Plant Disease* **93**, 440–451. doi:10.1094/PDIS-93-5-0440

Tripathi L, Tripathi JN, Kiggundu A, Korie S, Shotkoski F, Tushemereirwe WK (2014) Field trial of Xanthomonas wilt disease-resistant bananas in East Africa. *Nature Biotechnology* **32**, 868–870. doi:10.1038/nbt.3007

Tripathi L, Babirye A, Roderick H, Tripathi JN, Changa C, Urwin PE, Tushemereirwe WK, Coyne D, Atkinson HJ (2015) Field resistance of transgenic plantain to nematodes has potential for future African food security. *Scientific Reports* **5**, 8127. doi:10.1038/srep08127

Tripathi JN, Ntui VO, Ron M, Muiruri SK, Britt A, Tripathi L (2019) CRISPR/Cas9 editing of endogenous banana streak virus in the B genome of Musa spp. overcomes a major challenge in banana breeding. *Communications Biology* **2**, 46. doi:10.1038/s42003-019-0288-7

UNCST (2007) *The Biology of Bananas and Plantains.* Uganda National Council for Science and Technology, Kampala.

Waltz E (2014) Vitamin A super banana in human trials. *Nature Biotechnology* **32**, 857. doi:10.1038/nbt0914-857

Yadav J, Ogwok E, Wagaba H, Patil BL, Bagewadi B, Alicai T, Gaitán-Solis E, Taylor NJ, Fauquet CM (2011) RNAi-mediated resistance to Cassava brown streak Uganda virus in transgenic cassava. *Molecular Plant Pathology* **12**, 677–687. doi:10.1111/j.1364-3703.2010.00700.x

5

Learning from economists

A great source of information on the economics of GM crops is the annual conference of the International Consortium on Applied Bioeconomy Research (ICABR). As their webpage states, it is a 'unique, informal and international consortium of people interested in the bioeconomy, agricultural biotechnology, rural development and bio-based economy research' (www.icabr.org). Much of what appears in this chapter was obtained from people who have attended these meetings.

Why is genetic engineering accepted in medicine but not as readily in agriculture?

As mentioned in Chapter 1, the first bioengineered drug was commercialised in 1982 when the US Food and Drug Administration (FDA) approved recombinant human insulin. The first recombinant vaccine was approved in 1986 and by 2006 the FDA had approved more than 130 recombinant drugs and vaccines for human use. Similar approvals occurred in Europe, where the European Medicines Agency had approved 87 therapeutic proteins by 2006 (Paarlberg 2008).

Today that number would be much higher and, in addition to therapeutic proteins, recombinant DNA technology has been used to improve treatment strategies and to develop diagnostic kits, monitoring devices and new therapeutic approaches (Khan *et al.* 2016). In addition, several medically important proteins are being made in plants, an approach that will be discussed below as 'agroceuticals' made by 'pharming'.

In both the USA and Europe, genetic engineering (GE) as applied in medicine is supported far more than in agriculture (Priest *et al.* 2003). Why is this so? It cannot be due to poor understanding of the science, as drugs produced by GE are just as mysterious to the ordinary citizen as GM crops. Neither can it be antagonism towards corporate control, as the drugs are also produced by large, private sector companies. High product costs can also not be the issue because medicines made by GE are far more expensive than GM crops. Greater confidence in the regulatory agencies involved also cannot be the reason – just consider the cases of thalidomide and 'mad cow disease'.

In a 2016 paper published in the *Annual Review of Resource Economics*, Ronald Herring and Robert Paarlberg of Cornell University and the Harvard Kennedy School respectively (Herring and Paarlberg 2016), probed these questions. They concluded that drugs produced using GE receive greater support than GM crops partly because there is less chance of involuntary exposure to them as well as there being no environmental release. A person can control his or her exposure to a drug because it is prescribed by a medical doctor and always labelled. The first generation of foods derived from GM crops were not labelled, except, as described in Chapter 2, for the very early Flavr Savr tomatoes and the tomato paste they were made into in the UK. And they didn't last very long in the market. As will be discussed in Chapter 8, Europe began to require labels in 1997 on products whose content is derived from GM crops, but by then much of the damage had been done because the technology had been stigmatised and consumer resistance was high.

Another reason for the public acceptance of recombinant therapeutics is that drugs are produced inside tightly controlled laboratories, whereas GM crops are grown in the open, resulting in environmental exposure for humans, animals and other plants.

A further issue that Herring and Paarlberg discuss is the importance of benefits balancing risks (Herring and Paarlberg 2016). As drugs produced by biotechnology offer a range of benefits to users, these will usually be more important than the risks. In contrast, biotech crops, especially the first generation of *Bt* and herbicide-tolerant (HT) crops, had few visible benefits to ordinary citizens, especially in the relatively rich developed world. This is not necessarily so in the developing world where farmers and consumers are often the same people. However, as will be discussed in more detail in Chapters 9 and 10, opposition to GM crops is more complex than this.

The economics of GM crops worldwide

Early studies demonstrated that in the USA, nearly all of the economic gains from GM crops went to farmers, seed companies and patent holders. Indeed, only ~9–10% of the total economic advantage went to downstream purchasers (Falck-Zepeda *et al.* 2000). In more recent studies, Qaim (2009) and Bennett *et al.* (2013) have shown that larger shares of the benefit went to consumers as rates of adoption increased.

Peter Barfoot and Graham Brookes, both agricultural economists, started PG Economics, based in the UK, in 1999. Since 2005 they have been writing annual reports aimed at providing an accurate assessment of some of the key farm level economic effects associated with the adoption of GM crops worldwide. Their reports also try to facilitate more informed decision making in countries where GM crops are currently not permitted. The information covered in this section, on the economics of the most commonly grown GM crops, comes from Brookes and Barfoot (2018a).

Herbicide-tolerant crops

The main economic advantage of HT crops lies in less expensive and easier weed control. However, some growers have derived higher yields due to better weed control compared with that required for conventional crops. It must be stressed that these advantages are being partially offset in some regions where weeds have become resistant to glyphosate. Because glyphosate is a broad-spectrum post-emergence herbicide, it was often used as the only method of weed control by many farmers. This puts tremendous selection pressure on weeds, which helps the evolution of weed resistance to glyphosate and will be covered in Chapter 6.

On the positive side, and as will also be discussed further in Chapter 6, the use of HT crops can lead to decreased tillage or even to its absence, which can result in the shortening of the production cycle of crops such as soybeans. For instance, farmers in Argentina have been able to plant a crop of soybeans immediately after harvesting a wheat crop. The ability to harvest two crops in one growing season has added considerably to farm incomes and to the volume of soybeans that can be produced, both for local use and for export. For instance, HT soybeans have yielded average gross farm income benefits, in millions of US dollars, as follows: Argentina 18 567; Brazil 7220; Canada 200; USA 13 297.

HT canola, tolerant either to glyphosate or glufosinate, is grown in Canada, the USA and, more recently in Australia, while HT sugar beet tolerant to glyphosate is grown in the USA and Canada. In both these crops, the main benefit of higher farm income does, indeed, derive from yield gains arising from improved weed control compared with conventionally grown sugar beet.

Insect-resistant *Bt* crops

Increases in farm income due to *Bt* crops is mostly through the decrease in insect damage and the consequent higher yields. This is most evident in developing countries where insecticides are often prohibitively expensive. In addition, farmers, in these countries have gained from cost savings arising from reduced insecticide usage.

The average gross farm income benefits of *Bt* crops, in millions of US dollars, are as follows: Argentina 2392; Brazil 1832; Canada 185; USA 8450.

Global-level effects of GM crops

In 2016, the global gross farm income benefit due to GM crop use was US$18.2 billion. This is equivalent to adding 5.4% to the global value of the four main crops: soybeans, maize, canola and cotton. At a country level, unsurprisingly US farmers have been the largest beneficiaries because the adoption rate of all four crops between 1996 and 2016 was somewhat higher than 80%. Farm income gains in South America (mainly Argentina and Brazil, but also Bolivia, Colombia, Paraguay and Uruguay) were derived mainly from GM soybeans and maize. In China and India, on the other hand, farm income gains derive from *Bt* cotton.

In 2016, the cost to farmers for accessing GM technology was equal to 29% of the total value of technology gains. This is defined as the farm income (the gains referred to above) minus the cost of the technology payable to the seed supply chain. In developing countries, the total cost was equal to 20% of total technology gains compared with 36% in developed countries. Although circumstances vary between countries, the higher share of total technology gains accounted for by farm income in developing countries relative to developed countries reflects several factors. These include weaker provision and enforcement of intellectual property rights in developing countries as well as the higher average level of farm income gain per hectare derived by farmers in developing countries compared with those in developed countries.

Brookes and Barfoot (2018a) conclude their paper by referring to the fact that widespread use of GM crops has allowed farmers to grow more on less land. 'To maintain global production levels at 2016 levels, without biotech crops, would have required farmers to plant an additional 10.8 million hectares (ha) of soybeans, 8.2 million ha of maize, 2.9 million ha of cotton and 0.5 million ha of canola, an area equivalent to the combined land area of Bangladesh and Sri Lanka.'

They end on both cautionary and optimistic notes. The former regards the over-reliance on the use of glyphosate and the limited amount of crop and herbicide rotation by farmers in some regions. The latter includes the fact that there is a growing body of evidence in peer-reviewed literature of the positive effects of GM crop technologies. This explains why so many farmers around the world have adopted, and appear to be continuing to use, this technology.

What effects are GM crops having on human health?

In developed countries, crop spraying is highly mechanised, often depending on planes flying low over fields with the spray landing not only on the intended fields but also on neighbouring lands including residences. Many commercial farmers in South Africa have expressed their relief at not only the absence of noise pollution from these crop sprayers, but also the freedom from exposure to pesticides (Rosalind Steyn, pers. comm., 12 October 2019).

A different situation occurs in developing countries where farms and fields are much smaller, sometimes less than 1 ha. As a result, much of the field work, including weeding and spraying, is done by hand. Insecticide applications are often done using backpack sprayers, which can often result in absorption of the chemicals through the operator's skin. Pesticide poisoning as a result of this can be avoided by growing *Bt* crops, which require greatly reduced levels of insecticides. Reductions in farmer poisonings have been reported in South Africa (Bennett *et al.* 2003), China (Hossain *et al.* 2004), India (Kouser and Qaim 2011) and Pakistan (Kouser and Qaim 2013). Moreover, in a 2016 medical assessment of 246 Chinese farmers, Zhang *et al.* (2016) found that severe nerve damage was associated with insecticides used on non-*Bt* cotton.

Another health benefit of *Bt* maize, in particular, is a decrease in the incidence of certain types of cancer associated with high levels of dietary intake of fungal-infected maize. When non-*Bt* maize is stored, the small perforations made by insects become breeding sites for several fungi. Many of these produce toxins, known as mycotoxins, which can cause cancer. In a 21-year study of maize production, Pellegrino and colleagues (Pellegrino *et al.* 2018) found that *Bt* maize contained lower concentrations of mycotoxins of between 29% and 37%.

The cost savings due to these health benefits have been calculated for India and Pakistan. Kouser and Qaim (2011, 2013) estimated that in India, it is equivalent to US$14 million per year. They state this is a minimum amount as they did not include the positive spillover from *Bt* cotton, such as monthly expenditure on smoking, which has been positively correlated with poisoning incidence. It appears that farm workers who smoke during spraying are at a greater risk than non-smokers of inhaling toxic pesticide dust. These data agree with those found by Krishna and Qaim (2008). Estimates for such spillover effects suggest a further saving of US$9 million per year. Together, therefore, savings per year could reach considerably more than US$14 million.

In Pakistan, Kouser and Qaim (2013) took an unusual approach to determining the economic effects of *Bt* cotton. Besides the usual determination of financial benefits in terms of gross margins, they also quantified and monetised health and environmental benefits associated with the adoption of this technology. These benefits were associated with lower chemical pesticide use and included significant health advantages because of the lower incidence of acute pesticide poisoning. 'Farmers themselves value these positive effects at US$79 per acre, of which half is attributable to health and the other half to environmental improvements.'

What effect is GM technology having on the environment?

Brookes and Barfoot have written another paper that covers the effects that GM crops are having, positive and negative, on the environment (Brookes and Barfoot 2018b). In 2016, GM crops accounted for 48% of the global plantings of maize, soybeans, canola and cotton. Small areas of GM sugar beet have been grown in the USA and Canada since 2008, papaya in the USA since 1999 and in China since 2008, alfalfa, squash, potatoes and apples in the USA (the first mainly since 2011, and the other three since 2004, 2015 and 2016 respectively) and eggplants (brinjals) in Bangladesh since 2015.

The traits commercialised are tolerance to the herbicides, glyphosate and glufosinate, resistance to insect pests mainly due to *Bt* toxins, resistance to viruses, lack of browning in apples and low asparagine (resulting in low levels of acrylamide, a potential carcinogen) and reduced bruising in potatoes.

HT soybeans

From 1996 to 2016 there was a small net increase of 0.4% in the amount of herbicide active ingredient used on HT soybeans. This equates to ~13 million kg more over that time than would have been applied if non-HT soybeans had been grown. However, the authors also calculated the environmental effect as measured by the indicator, the environmental impact quotient (EIQ). Although this measurement is not perfect, these and other authors continue to use it because, in their view, it is a superior indicator to only using the amount of active ingredient used. They explain their reasoning in detail in the methods section of Brookes and Barfoot (2018b).

As the authors write: 'Assessing the full environmental impact of a pesticide use changes with different production systems. It is complex and requires an evaluation of risk exposure to pesticides at a site-specific level ... Undertaking such an exercise at a global level would require substantial and ongoing input of labour and time ... It is not surprising that no such exercise has, to date been undertaken, or likely to be in the near future.'

Thus, they use the EIQ indicator in conjunction with the changes in the volume of pesticide active ingredient applied. In the case of HT soybeans, there was a small net increase in the amount of herbicide active ingredient used (+0.4%), which is equal to ~13 million kg more than would have been applied to a conventional crop. However, when the environmental effect is measured by the EIQ indicator, there was a decrease of 13.4% between 1996 and 2016.

HT maize

In a similar study on HT maize, the authors found that the amount of active ingredient in herbicides for all countries had, in fact, decreased by 8.1% between 1996 and 2016, corresponding to an EIQ change of –12.5%. For both HT soybeans and HT maize the reductions were the greatest in the USA and Canada, compared with developing countries. This is because the former traditionally used higher amounts of herbicides.

HT cotton

The use of HT cotton resulted in a net reduction of the active ingredient of ~29 million kg over the 1996–2016 period. In terms of reduction in usage, this represents a value of 8.2%. In terms of the EIQ indicator, the savings amounted to a 10.7% net environmental improvement. In 2016 alone, the crop resulted in a 4 million kg reduction in the active ingredient, with an EIQ improvement of 16.7%.

Weed resistance

The problem of weed resistance is a common theme in discussions of all forms of agriculture, not just GM crops tolerant to specific herbicides. Although in the early years of GM HT crop adoption, some farmers used only glyphosate for weed control and contributed to the development of weeds becoming resistant to this herbicide, in the past 15 years more and more farmers have recognised the problem and diversified the herbicides that they apply. For example, in 2016, 89% of the HT soybean crop in the USA received one of the following active ingredients in addition to glyphosate: 2,4-D (used before planting), chlorimuron, fomesafen and sulfentrazone. This compares with 14% of the GM HT soybean crop receiving such treatments in 2006. These changes have influenced the mix, total amount, cost and overall profile of herbicides applied to GM HT crops and means that compared with the early 2000s, the amount and number of herbicide active ingredient used in most regions has increased, and the associated environmental profile, as measured by the EIQ indicator, has deteriorated. However, the amount of herbicide used on conventional crops has also increased over the same time period and, therefore in 2016, the environmental profile of GM HT crop use has continued to represent an improvement compared with the conventional alternative (as measured by the EIQ indicator; Brookes and Barfoot 2018b).

Insect-resistant maize and cotton

In many countries, planting of IR maize and cotton has reduced the use of insecticides quite remarkably. This is particularly evident in the case of cotton because conventional varieties have traditionally received intensive insecticide treatments, targeting bollworm and budworm pests. The savings have been less noteworthy in maize because the pests at which the *Bt* protein are targeted, such as stalk borers, tend to be less widespread than those in cotton. In addition, insecticides used on maize borers have limited effectiveness due to the fact that the borers are inside the stalks where the pesticide cannot penetrate. Thus, for instance, in the USA, before the advent of *Bt* maize, not more than 10% of the maize crop received insecticides targeting stalk-boring pests. Between 1996 and 2016, the reduction in the use on *Bt* maize has been 92 million kg, and for cotton it was an amazing 288 million kg. However, because of the complicated way in which the EIQ is measured the value for cotton is −32% and that for maize −59%.

Reduced fuel use

The savings in fuel associated with farmers making fewer spray runs in IR crops, as well as the switch from conventional tillage to reduced or no tillage in HT crops, have resulted in savings in CO_2 emissions. In 2016 alone, this amounted to a saving of 2945 million kg of CO_2, arising from reduced fuel use of 1309 million litres. These savings are equivalent to taking 1.8 million cars off the road for 1 year. The crop most responsible for these savings is HT soybeans in South America with a switch in tilling practices. Over the whole period of 1996–2016, the cumulative reduction in fuel use is equivalent to taking 16.7 million cars off the road.

The authors (Brookes and Barfoot 2018b) conclude by stating that currently more than 18 million farmers plant some form of GM seeds. These have helped them to be more efficient with their application of agents that protect their crops, which is clearly having a reduction in their environmental effect. This has been accompanied by a decrease in carbon footprint, allowing more carbon to be retained in the soil. Thus, together, IR and HT crops continue to provide net environmental benefits. It is also important to note that these findings are consistent with those reported by other authors (Fernandez-Cornejo *et al.* 2014; Klümper and Qaim 2014).

Although these data appear to end this chapter on a positive note, it should be borne in mind that the 18 million farmers planting GM seeds are made up mainly of the cotton farmers in India and China, not the producers of food such as the GM maize and soybeans grown in countries such as the USA, China, Argentina and Brazil. For nearly all food crops, including wheat, rice, most fruits and vegetables, GM approaches have been stopped by regulations and stigmatisation. These issues will be covered in greater detail in Chapters 9 and 10.

The next chapter will consider how to debunk many of the myths that have arisen regarding GM crops, partly due to misunderstandings (miscommunication) and partly due to deliberate distortion of facts (discommunication).

References

Bennett AB, Chi-Ham C, Barrows G, Sexton S, Zilberman D (2013) Agricultural biotechnology: economics, environment, ethics, and the future. *Annual Review of Environment and Resources* **38**, 249–279. doi:10.1146/annurev-environ-050912-124612

Bennett R, Buthelezi TJ, Ismael Y, Morse S (2003) *Bt* cotton, pesticides labour and health: a case study of smallholder farmers in the Makhathini Flats Republic of South Africa. *Outlook on Agriculture* **32**, 123–128. doi:10.5367/000000003101294361

Brookes G, Barfoot P (2018a) Farm income and production impacts of using GM crop technology 1996–2016. *GM Crops & Food* **9**, 59–89. doi:10.1080/21645698.2018.1464866

Brookes G, Barfoot P (2018b) Environmental impacts of genetically modified (GM) crop use 1996–2016: impacts on pesticide use and carbon emissions. *GM Crops & Food* **9**, 109–139. doi:10.1080/21645698.2018.1476792

Falck-Zepeda JB, Traxler G, Nelson R (2000) Surplus distribution from the introduction of a biotechnology innovation. *American Journal of Agricultural Economics* **82**, 360–369. doi:10.1111/0002-9092.00031

Fernandez-Cornejo J, Wechsler S, Livingston M, Mitchell L (2014) 'Genetically engineered crops in the United States'. USDA Economic Research Service Report Number 162. US Department of Agriculture, Economic Research Service. <www.ers.usda.gov>

Herring R, Paarlberg R (2016) The political economy of biotechnology. *Annual Review of Resource Economics* **8**, 397–416. doi:10.1146/annurev-resource-100815-095506

Hossain F, Pray C, Lu Y, Huang J, Fan C, Hu R (2004) Genetically modified cotton and farmers' health in China. *International Journal of Occupational and Environmental Health* **10**, 296–303. doi:10.1179/oeh.2004.10.3.296

Khan S, Ullah MW, Siuddique R, Habi G, Manan S, Yousaf M, Hou H (2016) Role of recombinant DNA technology to improve life. *International Journal of Genomics* **2016**, 2405954. doi:10.1155/2016/2405954

Klümper W, Qaim M (2014) A meta-analysis of the impacts of genetically modified crops. *PLoS One* **9**, e111629. doi:10.1371/journal.pone.0111629

Kouser S, Qaim M (2011) Impact of *Bt* cotton on pesticide poisoning in smallholder agriculture: a panel data analysis. *Ecological Economics* **70**, 2105–2113. doi:10.1016/j.ecolecon.2011.06.008

Kouser S, Qaim M (2013) Valuing financial, health, and environmental benefits of *Bt* cotton in Pakistan. *Agricultural Economics* **44**, 323–335. doi:10.1111/agec.12014

Krishna VV, Qaim M (2008) Potential impacts of *Bt* eggplant on economic surplus and farmers' health in India. *Agricultural Economics* **38**, 167–180. doi:10.1111/j.1574-0862.2008.00290.x

Paarlberg RL (2008) *Starved for Science: Keeping Biotechnology Out of Africa*. Harvard University Press, Cambridge, MA.

Pellegrino E, Bedini S, Nuti M, Ercoli L (2018) Impact of genetically engineered maize on agronomic, environmental and toxicological traits: a meta-analysis of 21 years of field data. *Scientific Reports* **8**, 1–12.

Priest SH, Bonfadelli H, Rusanen M (2003) The trust gap hypothesis: predicting support for biotechnology across national cultures as a function of trust in actors. *Risk Analysis* **23**, 751–766. doi:10.1111/1539-6924.00353

Qaim M (2009) The economics of genetically modified crops. *Annual Review of Resource Economics* **1**, 665–693. doi:10.1146/annurev.resource.050708.144203

Zhang C, Hu R, Huang J, Huang X, Shi G, Li Y, Yin Y, Chen Z (2016) Health effects of agricultural pesticide use in China: implications for the development of GM crops. *Scientific Reports* **6**, 1–8.

6

How to bust myths and the importance of communication

Over the years several myths have grown regarding GMOs and GM crops. Although it is important to use scientific facts to correct these myths, it is just as important to realise that people who believe them often do so because they reinforce some beliefs they hold. Therefore, it is not always helpful to simply give them the facts; it is just as necessary, possibly more so, to understand what lies behind their adherence to such myths. Thus, if a person is against the monopolisation of GM crops by multinationals, it is essential that it is against this background that facts are presented. In other words, it is necessary to gain the trust of the person involved before you can even begin to try to convince him or her to see GM crops in a different light.

The importance of trust

'Are public attitudes to new technologies driven more by our hearts or our heads? We'd like to think they are driven by our heads and that we all make decisions based on the scientific research and the data. *If only*!!' writes Craig Cormick (Cormick 2019). Cormick is an Australian science communicator and author, known for his writing on public attitudes towards GMOs and other new technologies. He goes on to cite a docudrama entitled *Brexit* in which the anti-Brexit campaigners aimed their messages at the public's head, while the pro-Brexiteers went for their hearts. Guess who won?

People tend to base their decisions on opinions and values and then look for facts that support these. If we wish to change their minds, we need to counter these in a way that is not immediately rejected. Thus, as mentioned above, when scientists are talking to an audience of people who may be neutral or vehemently against GM crops, it is imperative to gain their trust. It is important, from the word go, to explain what your expertise is and take responsibility for what you will say. If you take them into your confidence about what you are an expert in and where that ends, they are far more likely to trust what you have to say.

It is also good to state at the outset that there are aspects of GM crops that you find problematical. For instance, that monoculture, whether it is within a crop (only one or a few varieties planted) or between crops (only planting one type of crop), is not sustainable and can be harmful to the environment. Also, it would be better for consumers and farmers if GM crops were not in the hands of multinational companies only, but for that to happen the regulatory requirements would have to become more realistic. In the current regulatory environment, only the multinationals can afford to enter the market. In addition, it should be stressed that GM crops are not a magic bullet for solving food security issues but can be a part of the solution.

Another expert in this field is Kevin Folta, a pro-GMO professor in the Department of Horticultural Sciences at the University of Florida. In a paper describing one of his talks to an audience at the North Carolina State University, Jean Goodwin (2016) writes that Folta begins by arguing against some of the standard pro-GMO claims. For instance, he acknowledges that in many cases herbicide-tolerant crops have resulted in herbicide-resistant weeds, often due to monoculture. He also denounces the notion that GM crops could ever be the 'silver bullet' for sustainable agriculture. However, he did go on to say in defence of the technology:

> But I believe that there will be some important cases where this technology can change the [lives of] some people, some populations that can benefit, farmers in some countries can derive some sort of relief, farmers in Florida that can see their citrus crop restored. It may not change *the* world, but it can change *a* world, and if that world is your world, it's an important world.

If more scientists could take similar approaches, whereby they take time to understand their audiences' mindset, we might be able to correct at least some of the miscommunications. So, let's take a look at some of the misinformation that is being used, sometimes quite innocently because of a lack of knowledge or understanding, but too often for nefarious purposes, when it can be described as discommunication (i.e. deliberately trying to deceive using falsehoods). Let us now look at a few of the current myths and the ways in which they could be corrected.

Correcting myths

A common myth is that, since the advent of GM crops, farmers have not been able save their seeds. In fact, as discussed in Chapter 3, since the advent of hybrid seeds in the 1930s, farmers who plant them have had to buy seeds every year. This is because of the way in which hybrids are bred. Seed companies develop hybrids by breeding male and female lines that, when crossed, produce hybrid offspring with increased 'fitness', which produce better yields and make more money for the farmers. This is known as 'hybrid vigour', which is lost when the offspring seeds of that hybrid are planted (Fig. 6.1). As many GM crops are hybrids, if farmers plant their own seeds not only will they lose the hybrid vigour, they will also often lose the GM trait. This is because, during seed formation, the chromosomes of both parents are randomly inherited by the offspring. Thus, a quarter of the offspring will not carry the new gene.

A myth that is linked to this is that farmers are 'forced' to buy GM seeds. On the contrary, farmers are savvy people and they will buy the

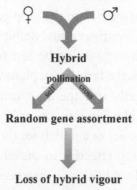

Fig. 6.1. The development of hybrid seed.

seeds that make them money. If these happen to be open-pollinated varieties, then those are the ones they will buy. If they happen to be GM, then they will buy those instead. In addition, in many public sector-funded projects, farmers are able to save and share GM seeds because no royalties are charged. Examples include the Hawaiian papaya (see Chapter 10), insect-resistant eggplants in Bangladesh (see Chapter 10) and, as discussed in Chapter 4, TELA maize varieties.

Another myth is that GM crops are a corporate plot to control developing nations and the world's food supply. The fact is that farmers in developing countries are growing more hectares of GM crops than those in developed countries (ISAAA 2018). They are doing so precisely because, as stated above, farmers buy the seeds that make them money. Planting GM seeds has also shown people in developing countries that these crops help to alleviate hunger.

But what about the environment? Many people think GM crops are harmful to the environment, whereas on the contrary, they have many environmental advantages. First, planting pest-resistant crops, such as *Bt* cotton, maize and eggplants, means that fewer pesticides need to be used. Not only does this reduce the use of chemicals, it also saves global greenhouse gas emissions by cutting down on the use of mechanical sprays, including the use of aerial spraying.

Second, herbicide-tolerant (HT) crops such as soybeans and maize, can be cultivated using less tilling. When farmers grow conventional varieties of these crops, they need to till the soil before planting, which allows weeds to grow. These then need to be killed with herbicides, most of which are not biodegradable. The farmer then must wait until the herbicide has dissipated before the crop can be planted, otherwise the plants will be killed by any residual herbicide remaining in the soil. During this time much topsoil can be lost from wind erosion. With HT maize and soybeans, the farmer can plant without, or with minimal tilling, because the weeds and the crop can grow together and the farmer can spray the field when it is convenient, killing only the weeds. The dead weeds can also act as a mulch for the crop. This process of no or minimal tilling is very effective in preventing the loss of valuable topsoil (Logan 2018). In addition, it has been estimated that, due to minimal tilling and decreased pesticide use, GM crops have reduced CO_2 emissions by 27 billion kg, equivalent to taking 16.7 million cars off the road for the period 1996 to 2016 (Brookes and Barfoot 2018).

Third, GM crops can increase each crop's biodiversity. It is relatively easy, far easier than with conventional breeding, to introduce one or a few genes into many different varieties. Indeed, it is estimated that 1167 *Bt* varieties of GM cotton were grown in India in 2014 and the number will be considerably greater by now (Choudhary and Gaur 2015).

There is, however, one negative environmental effect of HT crops and that is the overuse of Roundup®, which is leading to the build-up of weeds that are resistant to this herbicide. This is covered below under the section on the so-called 'superweeds'.

Another myth is the claim that GM crops are 'not natural' and that developers of them are 'playing God'. People have been selectively breeding plants and animals for countless years. Is a Poodle 'natural'? Is a seedless grape 'natural'? And in nature, bacteria are regularly transferring genes from themselves to different higher organisms (Lacroix and Citovsky 2016), so what is 'natural'? Agricultural improvement is, and always has been, unnatural. Conventional breeding, using inbred lines, developed over long periods, to produce hybrids is hardly 'natural' as shown in Fig. 6.1.

In addition, during the production of many new crops that are considered to be 'normal', mutations are induced using chemicals or irradiation. These agents are used when nature does not provide breeders with the trait they are looking for and one of the ways to find these traits is by creating mutations using DNA-damaging agents. The resulting plants are most certainly not 'natural'. New varieties derived by such methods are used worldwide and include rice, durum wheat, barley, soybean and legumes (Oladosu *et al.* 2016). In addition, millions of people regularly eat mutated grapefruit, apples or bananas (Schouten and Jacobsen 2007). Of course, none of these has had to undergo any type of regulation despite the agents that introduce the mutations being potentially carcinogenic.

Then there is the 'belief' that GM crops are a ploy by multinationals to sell more pesticides and herbicides, because why are so many GM crops sold by these corporations? The reason for this is that the cost of bringing a crop to market is so high that it is mainly the multinational companies that can afford to do this. I know this from my own experience when even the major international company, Pioneer, would not perform field trials on our MSV-resistant maize because they did not expect to be able to recover their costs when the only farmers who would need to buy the crop were African. As most African farmers are poor smallholders, Pioneer did not consider them able to pay the extra cost for the seed. This is also one of the reasons for the existence of the African Agricultural Technology Foundation, which accesses GM technology royalty free from multinationals for use by African farmers. In this way, at least some GM crops could be available more cheaply to these producers. Of course, as will be discussed later, this will only be possible if governments in Africa allow their farmers to obtain these products. At present this is only possible in South Africa, Sudan and, very recently, Nigeria and Eswatini (formerly Swaziland).

Regarding the expense involved, as long ago as 2011 Crop Life International published an article that estimated that the cost, both actual and in time involved, of discovery, development and compliance with regulations to bring a new plant trait to market was US$136 million (Phillips McDougall 2011). Of this, discovery accounted for slightly less than half of the expenses and complying with regulations and registration accounted for the rest. Considering that GM crops are the only crops to be subject to such stringent regulations, it is hardly surprising that the private sector is responsible for almost all that are available.

The Séralini affair

Another myth is the 'belief' that GM crops cause cancer due to the production of toxins. The origin of this myth was 19 September 2012, when Gilles-Éric Séralini, a professor of molecular biology at the University of Caen in France, published a peer-reviewed paper in the journal *Food and Chemical Toxicology* entitled 'Long term toxicity of a Roundup herbicide and a Roundup-tolerant genetically modified maize' (Séralini *et al.* 2012). The 2-year toxicity study, which cost €3.2 million, was funded by and run with the collaboration of the

Committee of Research and Independent Information on Genetic Engineering (CRIIGEN). Séralini is president of the Scientific Advisory Board of CRIIGEN, which opposes GM food. Séralini cofounded CRIIGEN in 1999 because he judged that studies on GM food safety were inadequate. As standard toxicological studies on rats are routinely done for 90 days, he decided that a 2-year study would be more appropriate.

Before 2012 Séralini had published other peer-reviewed papers that concluded there were health risks to GM foods and associated chemistries. For instance, in 2007 he and coworkers published a study funded by Greenpeace (Séralini *et al.* 2007) in which industry data were re-analysed using different statistical tests, revealing a miniscule effect of GM maize on rat organs. However, the European Food Safety Authority (EFSA 2007) concluded that the blood chemistry and organ weight values were within the normal range for control animals and they also stated that the paper had used incorrect statistical methods.

In the 2012 paper it was stated: 'In females, all treated groups died 2–3 times more than controls, and more rapidly. This difference was visible in 3 male groups fed GMOs. All results were hormone and sex dependent, and the pathological profiles were comparable' (Séralini *et al.* 2012).

Séralini held a press conference on the day the study was released to highlight his findings regarding the cancer his group had found in the rats. Selected journalists were given early access to the paper on condition they signed a confidentiality agreement, which prevented them from discussing the results with other scientists before the embargo expired. This is highly unusual, because most journals allow reporters to check their stories with independent experts, and was strongly criticised by, among others, the ethics committee of the French National Centre for Scientific Research, who called this approach inappropriate (Butler 2012). In addition, EFSA stated that the paper's conclusions were not supported by the data presented, stating: 'The design, reporting and analysis of the study, as outlined in the paper, are inadequate,' and that the paper was 'of insufficient scientific quality to be considered as valid for risk assessment'.

Soon after the publication of the article, *Food and Chemical Toxicology* began receiving letters to the Editor questioning its findings and some called for its retraction. The Editor-in-Chief requested, somewhat unusually, to see the raw data. He found no evidence of fraud but expressed concern regarding both the number of animals used in each study group (including controls) and the particular strain of rat that was used. As a result of the small sample size, the Editor-in-Chief concluded that no definitive conclusions could be reached regarding tumour incidence. The Sprague-Dawley rats used are known to have a high incidence of tumours and this normal variability could not be excluded as the cause of the higher incidence observed in the experimental rats. As a result, the data presented in the paper were inconclusive and the journal retracted the article 'Long term toxicity of a Roundup herbicide and a Roundup-tolerant genetically modified maize', which had been published in 2012 (Séralini *et al.* 2014a).

In June 2014, the original study was republished in the journal *Environmental Sciences Europe* with the addition of the entire dataset (Séralini *et al.* 2014b). The work was published without further review as the Editor considered that the previous reviews in a different journal were appropriate. In July the following year, the International Agency for Research on Cancer published a monograph on glyphosate, which contained an evaluation of the republished Séralini paper. They concluded that the study 'was inadequate for evaluation because the number of animals per group was small, the histopathological description of tumours was poor, and incidences of tumours for individual animals were not provided' (IARC 2015).

In 2012, around the time of the Séralini paper, a literature review was done on the value of multigenerational animal feeding trials (Snell *et al.* 2012). The authors examined 12 long-term studies of more than 90 days and up to 2 years, and 12 multigenerational studies, from 2 to 5 generations. They also referenced the 90-day feeding studies on GM food for which long-term or multigenerational study data were available. They stated that: 'If required, a 90-day feeding study performed in rodents, according to the OECD Test Guideline, is generally considered sufficient in order to evaluate the health effects of GM feed. The studies reviewed present evidence to show that GM plants are nutritionally equivalent to their non-GM counterparts and can be safely used in food and feed.'

In the section highlights they state:

- No sign of toxicity in the analysed parameters has been found in long-term studies.
- No sign of toxicity in the same parameters has been found in multigenerational studies.
- The 90-day OECD Guideline seems adequate for evaluating health effects of GM diets.

Despite the dubious nature of Séralini's paper, the idea that GM crops can cause cancer has become one of those myths that are hard to dispel. In my own city of Cape Town, the daily newspaper had the story on the front page, but after the article was retracted, made no mention of this at all. The power of miscommunication cannot be underestimated.

Immediately after the 2012 paper, Kenya banned further development or deployment of GM crops. This will be covered in Chapter 9.

Farmer suicides in India

Another myth that has been propagated mainly by the anti-GMO activist, Shiva Vandana (see also in Chapter 3), is that *Bt* cotton has been responsible for hundreds of farmer suicides in India (e.g. Shiva 2013, 2014). In the former article she stated: 'Monsanto has created [a] seed monopoly, crop monocultures and a context of debt, dependency and distress – which is driving the farmers' suicide epidemic in India. This systemic control has been intensified with *Bt* cotton. That is why most suicides are in the cotton belt. The highest acreage of *Bt* cotton is Maharashtra, and this is also where the highest farm suicides are.'

Is there a link between farmer suicides and *Bt* cotton? In an influential article, Keith Kloor, a freelance journalist in New York (Kloor 2014), quotes an Al Jazeera online story, which stated: 'More than 250,000 farmers have committed suicide in India after Monsanto's *Bt* cotton seeds largely failed. Many farmers decided to drink Monsanto pesticide, ending their lives.' He goes on to talk about a much-acclaimed

2011 documentary called *Bitter Seeds*, which chronicled this heartrending phenomenon and Monsanto's culpability. Indeed, 'Michael Pollan, a professor of journalism at the University of California, Berkeley, and author of the bestselling *Omnivore's Dilemma* and other food-related books, told his 300,000 followers on Twitter that *Bitter Seeds* was not to be missed, and lauded it as "a powerful documentary on farmer suicides and biotech seeds in India".'

As a result of this and many similar articles, Indian farmer suicides being due to Monsanto's *Bt* cotton has become a meme firmly embedded in many minds. However, there are several major problems with this belief. The first is the overwhelming popularity of this crop among Indian farmers, which, since its introduction in 2002, has been adopted by over 90% of Indian cotton farmers. In his article, Kloor cites India's Minister of Agriculture as saying that in 2012 the country harvested 5.1 million tons of cotton, well above the highest production of 3 million tons before the introduction of *Bt* cotton. As Ronald Herring, a political scientist at Cornell University, who has studied the seeming paradox and written on it extensively has stated: 'It is hard to imagine farmers spreading a technology that is literally killing them' (Herring 2009).

Not only does there seem to be no evidence that farmers using *Bt* cotton seed are more likely to commit suicide than others, but farmers that do use the seeds appear, on the whole, to be benefiting from them. Kathage and Qaim (2012) present data that show a 24% increase in cotton yield per acre through reduced pest damage and a 50% gain in profits among smallholder cotton farmers. Moreover, in a discussion paper published in 2008 by the International Food Policy Research Institute, the authors discredited the possibility of a direct causal or reciprocal relationship between *Bt* cotton and farmer suicides (Gruère *et al.* 2008).

Even in India the possibility of a link between *Bt* cotton and farmer suicides was challenged. In 2010, the Minister of Environment Affairs and Forestry, Jairam Ramesh, stated:

> *Bt*-cotton has catapulted India into second position in the world as far as cotton production is concerned … Over 90% of cotton farmers in India cultivate *Bt*-cotton. Studies done by the Tata Institute of Social Sciences, Mumbai challenges the popular NGO [non-government organization] belief that

there is a link between *Bt*-cotton and persistence of farmer suicides especially in Maharashtra' (Herring 2014).

What, then, could be causing these suicides among Indian farmers? In a paper entitled 'Political economy of suicide: financial reforms, credit crunches and farmer suicides in India', Sadanandan (2014) found that banking practices vary across India and that states with the highest incidence of farmer suicides were those that offered the least institutional credit to farmers. This resulted in smallholder farmers, whether they farmed cotton or not, falling into the hands of private lenders who charged exorbitant interest rates that could reach 45%. In fact, there was no direct correlation between cotton cultivation and a high incidence of farmer suicides.

But will these facts affect anti-GMO activists? In his paper, Kloor (2014) recounts that after attending a talk by Vandana Shiva he asked her about the mounting evidence that contradicted her suicide seeds claims. Apparently, she dismissed them saying that they were Monsanto studies, which, quite clearly, they were not. In fact, suicides are not higher among Indian farmers than among the general population. Kloor says the urban suicide problem is 'an untold and unwept story'. In an article for the *Free Press Journal* on 20 August 2012, S Ganesan wrote that not many people know that urban suicides outweigh farmer suicides by 7:1 (Ganesan 2012). Indeed, between 2007 and 2011, 87% of suicides were unrelated to farming, despite the fact that 70% of India's population live in rural areas.

In addition, to put Indian suicides into a global perspective, Japan, with a much higher gross national income than India, had a suicide rate ~250% higher than India's. In addition, the suicide rates in other wealthy countries such as the USA, Canada and Switzerland are all higher than in India (Ganesan 2012). One wonders about the not-so-hidden agenda of the anti-GMO lobby when they continue to try and link *Bt* cotton to the tragedy of Indian farmer suicides.

Terminator seeds

'Terminator seeds' was the name given by the anti-GMO lobbyists to a technology developed by Melvin Oliver, working for the United States Department of Agriculture (USDA) and with Delta and Pine

Land company, to create a cultivar that would be sterile on farmers' fields (Oliver and Velten 2001). The myth is that these GM cultivars were created to prevent farmers from saving seed and thus making them dependent on the multinational companies that develop them. In fact, the rationale behind the development of these sterile crops was to prevent the spread of pollen from fields of GM crops to neighbouring fields of non-GM crops. The technology came about as a result of activists campaigning against GMOs using, as one of their arguments, the possibility of pollen from GM crops 'contaminating' the fields of farmers nearby who did not wish to plant GM crops. One would think that rationally this development would have the approval of such activists, but this aspect of sterile seeds is seldom, if ever, mentioned.

The official name for this was genetic use restriction technologies and, under a research agreement with the USDA, the Delta and Pine Land company had exclusive rights to licence it to other parties (Lombardo 2014). However, because of public outcry, this technology has never been implemented in farmers' fields.

Superweeds

The term 'superweeds' is an emotive one aimed at inspiring fear. It is a term developed by the anti-GMO lobby to suggest that a weed is exceptionally strong or even invulnerable to any herbicide. In reality, 'superweeds' are no different from the herbicide-resistant weeds found in fields of conventional crops and which farmers have been battling for many years. Thus, weeds that are resistant to the herbicide Roundup can hardly be considered to be 'superweeds', which implies that they are resistant to all herbicides. This is analogous to 'superbugs', which are bacteria that are, indeed, resistant to all known antibiotics. These superbugs rightfully inspire fear, because of their potential to cause lethal, untreatable diseases in humans. By analogy, weeds resistant to one herbicide hardly warrant the term 'superweeds'.

That said, Roundup-resistant weeds are, indeed, a major and growing problem that needs to be addressed. However, to put this into context, herbicide-resistant weeds have been known since at least 1957 (Hilton 1957). Moreover, in a review of herbicide-resistant weeds, Heap

(2014) highlighted the fact that weeds have become resistant to 152 different herbicides and Westwood *et al.* (2018) emphasise the critical importance of managing weeds in a more integrated and sustainable manner. In the 1980s and 1990s weeds developed resistance to herbicides that inhibit the enzymes ALS (acetolactate synthase) and ACCase (acetyl-coenzyme A carboxylase). Currently, resistance to glyphosate, the active ingredient in Roundup, is becoming a major problem, although it has not yet surpassed the economic damage caused by the ALS and ACCase inhibitor resistant weeds (Heap 2014).

But in this situation, farmers are their own worst enemies – the more they continue to use Roundup without rotation, the greater chance there will be for the build-up of resistant weeds.

Glyphosate causes cancer – or does it?

Most HT crops that have been developed by genetic engineering (GE) are resistant to the herbicide Roundup, in which the active ingredient is glyphosate. Thus, by increasing the use of Roundup on crops, farmers and fieldworkers have greater exposure to glyphosate. This compound was classified by the International Agency for Research on Cancer (IARC) in 2015 as one that 'probably causes cancer in humans' (Guyton *et al.* 2015). A mere 8 months later EFSA found that the same compound was unlikely to pose a cancer risk to humans (EFSA 2015). Both claims have been extremely contentious, so which one is true? Or is there truth in both? And will farmers and fieldworkers spraying crops with Roundup have increased rates of cancer?

It would seem there is probably truth in both claims. The IARC is the cancer-research arm of the World Health Organization (a United Nations agency) and its report stated that there was 'mechanistic evidence', such as DNA damage to human cells due to exposure to glyphosate. This led them to classify the herbicide as 'probably carcinogenic'. EFSA, on the other hand, is concerned with whether there is sufficient evidence that a pesticide, when used according to the approved conditions, will pose a risk to human health. So, the former is hazard-based and the latter is risk-based. The position taken by EFSA has been echoed by a joint meeting of the FAO/WHO on pesticide residues (FAO/WHO 2016).

In 2019 a federal jury ordered Monsanto, the maker of Roundup, to pay US$80 million to a 70-year-old Californian man with cancer who had used the herbicide on his farm for three decades. The jury found that Roundup was a 'substantial factor' in his illness. This followed a similar decision, also by a Californian jury, in favour of a groundskeeper, Dewayne Johnson, awarding him US$78 million. Bayer AG, which recently bought Monsanto, said it would appeal both decisions.

In September 2019, three Californian medical associations filed an amicus brief supporting Bayer's appeal against the latter verdict. An amicus brief is a legal document filed in appellate court cases by non-litigants with a strong interest in the subject matter. In their brief, the associations made clear that they were not taking a side on the glyphosate/cancer issue. Instead, they criticised the way in which the case was handled (Barker 2019).

> Amici's point is that the answer to complex scientific questions such as that which the jury was required to resolve in this case should be based on accepted scientific evidence and rigorous scientific reasoning, not speculation and emotion.

It should also be noted that at least 15 regulatory and research agencies have conducted extensive long-term assessments of glyphosate, to determine whether, when used as directed, it increases the risk of certain cancers. Not one of these, including three at the WHO as well as the WHO itself, agreed with the highly controversial conclusion made by the IARC that glyphosate could cause harm to workers (Barker 2019).

This finding echoed that of Health Canada, which, having reviewed the two trial verdicts, stated that:

> After a thorough scientific review, we have concluded that the concerns raised by the objectors could not be scientifically supported when considering the entire body of relevant data. The objections raised did not create doubt or concern regarding the scientific basis for the 2017 re-evaluation decision for glyphosate. Therefore, the Department's final decision will stand ... No pesticide regulatory authority in the world currently considers glyphosate to be a cancer risk to humans at the levels at which humans are currently exposed.

This is clearly an ongoing saga that will continue for some time, but in the meantime users of glyphosate products should read the instructions on the label and follow them scrupulously.

With all these myths being spread, sometimes unwittingly, but often with malicious intent, what can scientists do to dispel them? One very useful organisation is the Alliance for Science.

The Alliance for Science

The Alliance for Science at Cornell University, directed by Sarah Evanega, is committed to furthering the cause of agricultural biotechnology by training and thus empowering science champions around the world with the tools and skills needed to communicate effectively. They provide accurate information based on science. They also share the stories of the people engaged in biotechnology through fact sheets, photographs, videos, blogs and other multimedia resources.

They hold a 12-week intensive training course for their Global Leadership Fellows Program on the Cornell University campus in Ithaca, NY, USA. This program enrols committed champions for agricultural biotechnology who are keen to become more effective communicators. The course is designed to equip them to provide science-based information in their home countries. It therefore tries to enrol small groups of fellows from the same country, so that, on their return home, they can work as a team, together with any former fellows who may also reside in their home countries.

The training, which is funded primarily by the Bill and Melinda Gates Foundation, focuses on strategic planning, grassroots organising and innovative communications with a strong focus on stories around agricultural innovation. Throughout the course, fellows work to build strategic communications plans for implementation in their local contexts.

As stated on their webpage (www.allianceforscience.cornell.edu): 'Upon completion of this certificate program, Fellows become members of an international professional network and cohort uniquely qualified to promote evidence-based decision-making around global issues such as food security, agricultural development, environmental sustainability, and climate change.'

One example of the work is Science Stories Africa, the brainchild of a fellow from Uganda, Patricia Nanteza. She is using Uganda's National

Theatre to shift from the usual fiction stories into fact-packed tales told by local scientists themselves. The event, the first of its kind in the theatre's 60-year history, was created to provide a platform for African scientists to tell their stories of how technology offers solutions to African challenges.

To quote Nanteza from the website: 'The West is the leader in content creation, and Africa tends to simply consume and believe whatever narrative comes from there. This has been evident in the GMO debate, with some African nations saying no to GE crops simply because a very vocal section of the West is against it.'

When Nanteza became aware of this 'opinion damping,' she wanted to find a way in which '[she] could contribute to changing the narrative around technology adoption, especially in agriculture. That is when I got the idea of Science Stories Africa. We want to enable the public to appreciate the innovations and gradually adopt them.'

Nanteza has local scientists talking about how they have developed GE bananas to resist bacterial wilt and nematodes and to produce vitamin A, including their frustrations when initial attempts failed. In this way the audiences become engaged in the real-life experiences of local scientists with whom they can identify. All the scientists who have told their stories were inspired by the urgent need to solve problems in their communities. Their compelling stories would probably have remained within the laboratories or in obscure scientific journals if it were not for the novel concept of Science Stories Africa. Now they can be used to entertain and educate ordinary people who would normally come to the theatre to enjoy an evening of fiction.

Similarly, a Nigerian Alliance for Science fellow, Modesta Abugu, has formed a team with other fellows and interested stakeholders, including the Open Forum on Agricultural Biotechnology in Africa (OFAB) to canvass for the passage of a biosafety bill that could operate under the *National Biosafety Management Agency (NBMA) Act* of 2015. In an appraisal of this Act (Muzan 2018) the author wrote that it 'reinforces the institutional structure with the aim of ensuring the effective management of all components of the Nation's biosafety.' It could be the 'national authority on biosafety in Nigeria to effectively exploit the potential benefits of modern biotechnology and efficiently guard against associated risks.' However, at that stage a biosafety bill still needed to be passed in parliament in order to allow for the commercialisation of GM crops under research.

Abugu and her team at OFAB Nigeria, under the leadership of Dr Rose Gidado, saw a gap in the way science communicators understood and communicated about GMOs, using lots of facts that usually left people more confused and increasingly sceptical. Based on the experiences they had gained from the Alliance for Science programme, they designed a training programme for over 100 science communicators from different parts of the country on the science and benefits of GM crops. They used a grassroots model that they had learned from the Alliance whereby their communication was not only based on facts, but also on shared values. This helped to shape people's opinions and improve their understanding of how GM crops could benefit their lives. Their aim is to take the message to farmers, who, in turn, would place demands on the policy makers to pass the bill into law. 'Replicating the communication model from the Alliance was a game changer', Modesta said.

Journalists were not left out of this venture. Realising that most people have little or no interest in engaging in science-based conversations, two fellows from the Nigerian Alliance for Science team, Opuah Abeikwen and Micheal Etta, set up a program called the 'Science Café'. This is an informal information-sharing platform aimed at building science literacy, advancing innovation and overcoming public fears about biotechnology. The Café sessions, which are held monthly, provide an opportunity for stakeholders and journalists to engage with each other through meaningful debates that are based on sound scientific principles and promote information sharing.

'It is unfortunate that the work of scientists ends up on shelves and consumers do not reap the benefits of the research being conducted by our research institutes on agricultural biotechnology', Opuah Abeikwen said. 'The idea behind Science Café Nigeria is to be a voice for safe science in the country. So far, the science hangouts have provided an informal opportunity for journalists to ask pertinent questions on GM crops in order to enhance their understanding and reporting of the technology.'

The team has been successful in their mission because the Nigerian government has ruled that the NBMA will be the agency that approves confined field trials and commercial release of GM crops. For instance, the approval for commercial release of *Bt* cowpea was issued by the NBMA on 22 January 2019. In a statement by the executive director of the Institute for Agricultural Research at Ahmadu Bello University in Zaria, northern Nigeria, Professor Ibrahim Abubakar stated, 'Cowpea is the most important food grain legume in Nigeria. The low yield of the crop in Nigeria is due to many constraints, particularly pod boring insects, which cause up to 90 percent yield loss in severe infestation cases.' Other GM crops that are currently undergoing field trials could also be approved by the NBMA.

Scientists in Kenya are following the example set by Nigerians in establishing Science Cafés. At one entitled 'Bridging the Gap between Scientists and Journalists', the scientists present were urged to reach out to journalists and editors through the Kenya Editors' Guild. This could help deepen the understanding of media personnel regarding the problems between research into GM crops and their uptake by farmers (ISAAA 2019).

Ken Monjero, more commonly known as 'Ken from Kenya' is engaging youth through his education campaign. Ken is the force behind Kenya's first informal science education centre, based at the Kenya Agriculture and Livestock Research Organization's centre for biotechnology, where he feeds children's curiosity for science and inspires them to pursue science. It is in this context that he is also able to help children (and therefore their parents) navigate misinformation on biotechnology and other misunderstood areas of science.

In 2019, a three-woman team of young farmers hailing from Ghana, Zimbabwe, and Zambia formed a campaign to organise and empower 'Women Who Farm,' under that banner. This group of energised and youthful female farmers are working not only to ensure that women farmers have access to biotechnology, but to other life-improving technologies as well.

Beyond Africa, Arif Hossain, a medical doctor, has, since his Alliance for Science fellowship and follow-on training, built a similar organisation in Bangladesh. Farming Future Bangladesh (FFB) seeks to engage stakeholder groups who have a tremendous amount to gain from biotech crops in the pipeline in Bangladesh, such as Golden Rice and *Bt* eggplant (also known as brinjal and talong; see Chapter 10). His organisation works with the nutrition community, to engage with them on the benefits of these crops. FFB also helps the media and the faith community engage with scientists and other science-based resources.

Finally, Pablo Orozco, a 2016 Global Leadership Fellow from Guatemala has launched a global campaign to help promote evidence-based decision-making around the meetings on the Convention on Biodiversity. He has rallied fellows from across cohorts and countries to stand in support of biotech science at the global level.

Through organisations such as the Alliance for Science and its various off-shoots, there is an increasing number of scientists and communicators attempting to bust the many myths that surround GM crops. The Alliance's website carries a whole section on myths and realities. As we have seen, some myths do have an element of truth in them, but others are pure fantasy. It is up to scientists and science communicators to work harder to ensure that the truth prevails. Perhaps, through their efforts, more countries in the developing world will see the benefits that could be afforded by these crops and allow their farmers to at least test for themselves whether they work or not. In the next chapter we look at countries that got it right, in allowing their farmers such access and why they did so.

References

Barker T (2019) 'Trials should be settled by 'scientific evidence, not speculation and emotion'. Genetic Literacy Project, 16 October 2019. <https://geneticliteracyproject.org/2019/10/16/trials-should-be-settled-by-scientific-evidence-not-speculation-and-emotion-in-unusual-twist-california-medical-groups-join-appeal-of-jury-verdict-finding-monsantos-round/>

Brookes G, Barfoot P (2018) Environmental impacts of genetically modified (GM) crop use 1996–2016: impacts on pesticide use and carbon emissions. *GM Crops & Food* **9**, 109–139. doi:10.1080/21645698.2018.1476792

Butler D (2012) Hyped GM maize study faces growing scrutiny. *Nature* **490**, 158. doi:10.1038/490158a

Choudhary B, Gaur K (2015) *Biotech Cotton in India, 2002 to 2014*. ISAAA Series of Biotech Crop Profiles. ISAAA, Ithaca, NY.

Cormick C (2019) Public attitudes toward new technologies: our post-truth, post-trust, post-expert world demands a deeper understanding of the factors that drive public attitudes. *Science Progress* **102**, 161–170. doi:10.1177/0036850419851350

EFSA (European Food Safety Authority) (2007) EFSA review of statistical analyses conducted for the assessment of the MON 863 90-day rat feeding study. European Food Safety Authority.

EFSA (European Food Safety Authority) (2015) Peer review of the pesticide risk assessment of the potential endocrine disrupting properties of glyphosate. *EFSA Journal* **13**, 4979–4999.

FAO/WHO (2016) 'Pesticide residues in food'. 16 May 2016. <http://www.fao.org/agriculture/crops/thematic-sitemap/theme/pests/jmpr/jmpr-rep/en/>

Ganesan S (2012) 'Bare facts about farmer suicides'. *The Free Press Journal*, 20 August 2012, India.

Goodwin J (2016) 'Demonstrating objectivity in controversial science communication: a case study of GMO scientist Kevin Folta'. OSSA Conference Archive 69. <https://jeangoodwin.files.wordpress.com/2018/10/goodwin-folta-ossa-2016.pdf>

Gruère GP, Mehta-Bhatt P, Sengupta D (2008) '*Bt* cotton and farmer suicides in India: reviewing the evidence'. IFPRI Discussion Paper 00808. International Food Policy research Institute, October 2008.

Guyton KZ, Loomis D, Grosse Y, El Ghissassi F, Benbrahim-Tallaa L, Guha N, Scoccianti C, Mattock H, Straif K (2015) Carcinogenicity of tetrachlorvinphos, parathion, malathion, diazinon, and glyphosate. *The Lancet Oncology* **16**, 490–491. doi:10.1016/S1470-2045(15)70134-8

Heap I (2014) Herbicide resistant weeds. In *Integrated Pest Management*. (Eds D Pimentel and R Peshin) pp. 281–301. Springer, Dordrecht.

Herring R (2009) Persistent narratives: why is the "failure of *Bt* cotton in India" story still with us? *AgBioForum* **12**, 14–22.

Herring RJ (2014) State science, risk and agricultural biotechnology: *Bt* cotton to *Bt* Brinjal in India. *The Journal of Peasant Studies* **42**, 159–186. doi:10.1080/03066150.2014.951835

Hilton HW (1957) *Herbicide tolerant strains of weeds*. Hawaiian Sugar Plant Association Annual Report 69. Springer, Switzerland.

IARC (2015) *Glyphosate*. IARC Monographs – 112. p. 35. <https://monographs.iarc.fr/wp-content/uploads/2018/06/mono112-10.pdf>

ISAAA (2018) 'Global status of commercialized biotech/GM crops in 2018: biotech crops continue to help meet the challenges of increased population and climate change'. ISAAA Brief No. 54. ISAAA, Ithaca, NY.

ISAAA (2019) 'Engaging editors crucial for improved science reporting in Kenya'. Crop Biotech Update, 10 October 2019. <www.isaaa.org/kc/cropbiotechupdate/article/default.asp?ID=17776>

Kathage J, Qaim M (2012) Economic impacts and impact dynamics of Bt (*Bacillus thuringiensis*) cotton in India. *Proceedings of the National Academy*

of Sciences of the United States of America **109**, 11652–11656. doi:10.1073/
pnas.1203647109

Kloor K (2014) The GMO-suicide myth. *Issues in Science and Technology* **30**(2),
65–78.

Lacroix B, Citovsky V (2016) Transfer of DNA from bacteria to eukaryotes.
mBio **7**(4), e00863-16. doi:10.1128/mBio.00863-16

Logan TJ (2018) *Effects of Conservation Tillage on Ground Water Quality: Nitrates
and Pesticides.* CRC Press, Boca Raton, FL.

Lombardo L (2014) Genetic use restriction technologies: a review. *Plant
Biotechnology Journal* **12**, 995–1005. doi:10.1111/pbi.12242

Phillips McDougall (2011) 'The cost and time involved in the discovery,
development and authorisation of a new plant biotechnology derived trait'.
A Consultancy Study for Crop Life International September 2011. Phillips
McDougall, UK.

Muzan MA (2018) Institutional mechanisms for biosafety in Nigeria: an
appraisal of the legal regime under the National Biosafety Management
Agency Act, 2015. *Law, Environment and Development Journal* **14**, 29–44.
<http://www.lead-journal.org/content/18029.pdf>

Oladosu Y, Mohd Y, Rafii NA, Ghazali H, Asfaliza R, Harun AR, Gous M,
Magaji U (2016) Principle and application of plant mutagenesis in crop
improvement: a review. *Biotechnology, Biotechnological Equipment* **30**, 1–16.
doi:10.1080/13102818.2015.1087333

Oliver MJ, Velten J (2001) Development of a genetically based seed technology
protection system. In *Dealing with Genetically Modified Crops.* (Eds RF
Wilson, CT Hou and DF Hildebrand) pp. 110–114. AOCS Press,
Champaign, IL.

Sadanandan A (2014) Political economy of suicide: financial reforms, credit
crunches and farmer suicides in India. *Journal of Developing Areas* **48**,
287–307. doi:10.1353/jda.2014.0065

Schouten HJ, Jacobsen E (2007) Are mutations in genetically modified plants
dangerous? *Journal of Biomedicine & Biotechnology* **2007**, 82612.
doi:10.1155/2007/82612

Séralini GE, Cellier D, de Vendomois JS (2007) New analysis of a rat feeding
study with a genetically modified maize reveals signs of hepatorenal
toxicity. *Archives of Environmental Contamination and Toxicology* **52**,
596–602. doi:10.1007/s00244-006-0149-5

Séralini G-E, Clair E, Mesnage R, Gress S, Defarge N, Malatesta M, Hennequi
D, de Vendômois JS (2012) RETRACTED: Long term toxicity of a
Roundup herbicide and a Roundup-tolerant genetically modified maize.
Food and Chemical Toxicology **50**, 4221–4231. doi:10.1016/j.
fct.2012.08.005

Séralini G-E, Clair E, Mesnage R, Gress S, Defarge N, Malatesta M, Hennequi D, de Vendômois JS (2014a) Retraction notice to 'Long term toxicity of a Roundup herbicide and a Roundup-tolerant genetically modified maize' [Food Chem Toxicol (2012) 50, 4221–4231]. *Food and Chemical Toxicology* **63**, 244. doi:10.1016/j.fct.2013.11.047

Séralini G-E, Clair E, Mesnage R, Gress S, Defarge N, Malatesta M, Hennequin D, de Vendômois JS (2014b) Republished study: long-term toxicity of a Roundup herbicide and a Roundup-tolerant genetically modified maize. *Environmental Sciences Europe* **26**, 14–31. doi:10.1186/s12302-014-0014-5

Shiva V (2013) 'Seeds of suicide and slavery versus seeds of life and freedom'. Opinion: Business & Economy. <http://www.aljazeera.com/indepth/opini on/2013/03/201332813553729250.html>

Shiva V (2014) 'The seeds of suicide: how Monsanto destroys farming'. <http://www.globalresearch.ca/the-seeds-of-suicide-how-monsanto-destroysfarming/5329947>

Snell C, Bernheim A, Berge J-B, Kuntz M, Pascal G, Paris A, Ricroch AE (2012) Assessment of the health impact of GM plant diets in long-term and multigenerational animal feeding trials: a literature review. *Food and Chemical Toxicology* **50**, 1134–1148. doi:10.1016/j.fct.2011.11.048

Westwood JH, Charudattan R, Duke SO, Fennimore SA, Marrone P, Slaughter DC, Swanton C, Zollinger R (2018) Weed management in 2050: perspectives on the future of weed science. *Weed Science* **66**, 275–285. doi:10.1017/wsc.2017.78

Countries that got it right and why

What do I mean by a country 'getting it right'? Such a country needs to have a government that is supportive of innovations and new technologies that improve agricultural production and make the lives of farmers more profitable and less stressful in the event of losses in productivity. It should also have a regulatory system in place that is flexible, operates on a case-by-case system and makes science-based decisions. The regulations should not act as barriers to the development and implementation of GM crops. And finally, the government should encourage private enterprises to develop and commercialise such crops.

In this light, let us consider four countries that qualify under these criteria. Not all of them have fulfilled every one of the criteria, but they have done so sufficiently to enable the majority of farmers in their country to grow and benefit from GM crops.

South Africa

South African farmers started planting GM crops commercially as early as 1998. Thus, even in those early days the government was supporting these innovative technologies that enabled farmers to become more profitable. In 2018 they were growing 2 million hectares of GM maize, 0.7 million of soybeans and 37 000 of cotton. This means that ~85% of maize farmers chose to plant GM maize in that year, as some will alternate between GM and non-GM seeds depending on personal requirements such as varietal preferences, intended use of the crop etc. (ISAAA 2018). Most (66%) of the GM maize varieties carry both the *Bt* and herbicide-tolerant (HT) traits, commonly referred to as stacked traits. The fall army worm, mentioned in Chapter

4, which has devastated many of the maize fields in other parts of Africa, had much less of an impact here, which has been attributed to the high percentage of *Bt* maize that was planted. In addition, over the years the country has been planting GM maize the higher yields have meant there is less need to plant crops on marginal land. Indeed, the maize yield has doubled since 1998 (ISAAA 2017b).

There has been a problem, however, with GM soybeans because farmers were keeping significant amounts of their harvested seeds for replanting the next season. Why is this a problem? Plant breeders invest significant sums of money developing and testing GM crops. It is not unusual for a new variety to cost millions of US dollars to bring to commercialisation. These investments need to be recovered by the sale of seeds. Farmers who plant hybrid crops buy seeds every year because saved seeds lose the qualities and traits developed by the breeders. Farmers regularly calculate whether the cost of improved seeds is worthwhile or whether they will make similar profits from buying conventional seeds of a lower quality. Although this is the case for maize, whether GM or not, it is not so for soybeans, most of which are not hybrids. Thus, farmers who plant saved soybean seeds are preventing the breeders from recovering their 'innovation' costs. If breeders are unable to recoup these costs they will not be able to afford to continue to develop new beneficial varieties, which harms both them and the farmers.

Several farmer organisations, including Grain SA and SANSOR (South African National Seed Organization), formed the South African Cultivar and Technology Agency and came to an agreement regarding levies. Included in the agreement was that 20% of the levies would go to help the transformation of black farmers, who, under the apartheid system of government, had been severely disadvantaged in terms of land acquisition, access to financial aid, freedom of movement and many other opportunities to improve their livelihood. The 20% of the levies would go towards helping to right some of these wrongs and to assist these soybean farmers to become more financially independent (ISAAA 2017b). However, as will be seen later, this does not help the smallholder maize farmers who greatly outnumber soybean farmers.

As mentioned above, one of the reasons why South Africa 'got it right' was because it was an early adopter of GM crops. Why was this the case? One of the reasons is that the country had been dealing with GMOs since 1978 when the Council for Scientific and Industrial Research (CSIR), then the major government research funding agency, started the South African Committee for Genetic Experimentation (SAGENE). The CSIR had started to receive research grant proposals for work on GM bacteria and it was decided that guidelines along the lines of those developed in the USA by the National Institutes of Health, the NIH Guidelines for Research into Recombinant DNA (see Chapter 1), were needed.

SAGENE fulfilled an excellent development role because it required universities to improve laboratory standards in order for their academic staff to obtain research grants for work on GMOs. They also ran training programmes that helped to bring many young scientists into this area of research. After some years SAGENE had accomplished much of what it was set up to achieve and therefore went into abeyance, meeting only once or twice a year. That, however, changed in the 1990s.

In 1990, SAGENE received an application from the multinational company Calgene Inc. for field trials of GM cotton resistant to the herbicide bromoxynil (BXM™). Shortly thereafter they received an application from Clark Cotton to conduct trials of Bollgard® cotton seed, carrying a *Bt* gene. Faced with these applications, the South African government officially reconstituted SAGENE, announcing it in the *Government Gazette* of 15 May 1992 (Thomson 2013).

SAGENE then drew up procedures for assessing whether to approve applications for importation, trial release or general release of GM crops. However, it was not a statutory body of the government and therefore could not impose penalties for any infringements of its procedures. As a result, the government set about developing a *Genetically Modified Organisms Act*, which was published on 23 May 1997. However, in true government style, it took a considerable time before the Act could be implemented, because the regulations had to be approved by parliament. As a result, SAGENE took on an interim role until this took place in December 1999. In the interim, SAGENE handled 27 applications, of which 13 were for maize, 4 for cotton, 2 for soybeans, 1 each for canola, strawberries, eucalyptus and apple and 4 for microorganisms. The traits included insect resistance, fungal and viral resistance and herbicide tolerance.

As a result, by the time the GMO Act came into being, the government was reasonably familiar with GM crops. It also had advisors from the scientific and agricultural sectors. One of these advisory bodies was, and still is, the National Advisory Committee on Innovation, which advises the Department of Science and Innovation. Together they drew up the National Biotechnology Strategy in 2001, which resulted in several Biotechnology Regional Innovation Centres, including the National Innovation Centre for Plant Biotechnology (PlantBio). In 2013 they produced the Bio-Economy Strategy, which aimed to bring innovative research to commercialisation. The encouragement of private enterprise to become involved in the mobilisation of GM crops for public use is another feature of countries that got it right.

Another reason why South Africa became an early acceptor of GM crops was the organisation AfricaBio. This was started in 1990 and became a registered Section 21 company in 2000 (Thomson 2013). According to their website they are 'an independent, non-profit biotechnology stakeholders' association for the safe, ethical, and responsible research, development, and application of biotechnology and its products' (www.africabio.org). In the early years of its existence it played a pivotal role in informing farmers, commercial and smallholder alike, about the role that GM crops could play in agriculture. They ran workshops to teach them about the sustainable use of both *Bt* and HT crops to prevent the build-up of resistance in either insects against the *Bt* toxin or in weeds against the relevant herbicides.

A third reason for the early acceptance of GM crops was that the great majority of commodity crop production is carried out on modern, commercial farms where the benefits of the GM input traits were apparent. By making GM seeds available to such farmers as early as 1998, the South African government certainly 'got it right'. In addition, by facilitating, more recently, the payment of levies to breeders by soybean farmers, it is enabling breeders to continue to bring novel varieties to market. This could be used as a model for other countries facing similar problems.

But what about the smallholder, mainly subsistence, farmers? Unfortunately, in this respect, the government has not yet 'got it right'. Why is this? South Africa is the only country where the majority of its citizens eat GM white maize up to three times a day. Yellow maize is used for animal feed. In 2016, Marnus Gouse and colleagues carried out a study to determine what effect growing GM maize had on smallholder farmers in a region where this is grown extensively (Gouse *et al.* 2016). They stated that 'although there is an extensive and growing body of literature on the economic impact of the adoption of GM crops in both developing and developed economies, there is only scant evidence that the technology has had any specific and distinguishable impact among female and male [smallholder] farmers.'

In order to provide evidence of the role that GM maize plays in the lives of smallholder farmers in South Africa, Gouse and his colleagues studied gender-specific adoption and performance of *Bt* and HT maize in the KwaZulu-Natal province. They chose this region because many of the smallholder farmers there are women. Their results, both quantitative and qualitative, showed that women farmers, traditionally responsible (together with their children) for weeding, valued the labour-saving advantages of HT maize. Labour demand is greatest during the crucial period when the land is prepared, crops are planted and weeded. Labour-saving advantages are amplified by the trend of migration away from rural areas to cities. Among male farmers, higher yields were the main reason behind adoption (Gouse *et al.* 2016).

Another study looked specifically at the use of *Bt* maize by smallholder farmers. Kotey *et al.* (2016) found a problem linked to the possible build-up of insects that become resistant to the *Bt* toxin. Farmers need to plant a buffer zone of non-*Bt* plants around their *Bt* crop to maintain a population of *Bt*-susceptible insects that can reproduce with any resistant insects that develop and thus sustain *Bt*-susceptible individuals in the pest population. The problem they identified was the lack of good extension and advisory support received by these farmers, especially in the poorer areas of the country. Smallholder *Bt* maize farmers need to be made aware of the stewardship requirements of the buffer zone and this information must be transferred to them in a manner that is understandable. If this is not done, the long-term sustainability of this technology is at risk. The study showed that the extension personnel in the regions they studied lacked adequate training to effectively disseminate GM maize technology to those farmers.

The authors (Kotey *et al.* 2016) end by recommending that the Department of Rural Development and Agrarian Reform (DRDAR) train the extension personnel on good stewardship methods and the consequences of non-compliance. In addition, the DRDAR should make non-*Bt* hybrids available to the farmers in order to allow them to use such plants as buffers to prevent insect resistance from developing. If this is not done, the country's commercial farmers will have 'got it right', but their smallholder counterparts will not.

But how does the South African public respond to GM crops? The Human Sciences Research Council conducted a nationwide survey in 2017 to determine public perceptions of GM crops (Gastrow *et al.* 2018). Some of the findings were as follows.

- The South African public is less informed, but more positive about biotechnology, and specifically GM food, than Europeans. For instance, 53% compared with 31% believe that it is safe to eat.
- Public awareness of biotechnology increased between 2004 and 2015. Is this due to increased levels of information, possibly due to increases in labelling?
- Younger people, with higher levels of education and higher living standards, are more positive about biotechnology.
- Most South Africans believe that GM foods are good for the economy and the overall risk/benefit assessment is positive, although only about half of the respondents engaged with these questions.

Is it possible to explain these attitudes which, on the whole, are more positive towards GM crops than those of Europeans? One possible reason is that, unlike Europeans, South Africans have been eating GM food, such as maize, and food derived from GM crops, since the early 2000s. Although many have only become aware of this somewhat recently, it could possibly have swayed their opinions knowing that this has been happening without harmful human effects. Gastrow *et al.* (2018) do not offer any opinions, but state that factors such as these 'pave the way for strategic interventions that will build up public knowledge, while at the same time cultivating constructive engagement between the public and the biotechnology sector'.

When considering how South Africa 'got it right', it is clear that the government supported this technology as early as 1998 as it could see the benefits it brought to commercial farmers, both in terms of increased profits but also in rendering their lives less stressful due to insect pests and weed problems. The regulatory system has been flexible with each variety being treated on a case-by-case basis. And finally, it has, albeit only fairly recently, put systems in place to assist the commercialisation of GM crops developed within the country.

Despite this, several challenges still remain. One is based on the experience of HT crop farmers in the USA: the build-up of weeds resistant to the active ingredient glyphosate present in RoundUp and similar herbicides (Green 2018). This is due to the excessive use of these herbicides, similar to the way in which the overuse of antibiotics leads to antibiotic resistance in humans and some livestock. Currently in South Africa the only glyphosate-resistant weeds are not associated with GM crops (International Survey of Herbicide Resistant Weeds 2019). However, that does not mean that the country can ignore the potential for the development of such weeds in fields of GM crops. The message that problems can result from the overuse of a single herbicide needs to be conveyed to farmers.

Another problem lies with the use of GM crops by smallholder farmers. There needs to be an improvement in information transfer by extension personnel regarding, as mentioned above, the importance of correct stewardship, both the requirement of a buffer zone with *Bt* crops and the avoidance of over-exposure of HT crops to glyphosate-containing herbicides.

In addition, it is essential to continue with the education of the public on the advantages of GM crops to agricultural productivity, but also to stress the responsibilities that come with their cultivation. The advantages might become more obvious in the future as climate change leads to increasing problems with drought and the drought-resistant crops, mentioned in Chapter 4, become a reality.

Canada

> Canada regulates products derived from biotechnology processes as part of its existing regulatory framework for "novel products." The focus is on the traits expressed in the products and not on the method used to introduce those traits ... Advertising or labeling the presence of GMOs in particular food is voluntary unless there is a health or safety concern.

Thus reads the introduction to the Law Library of Congress's (2012) article entitled 'Restrictions on Genetically Modified Organisms: Canada'. Some scholars have noted that Canada generally espouses a permissive attitude towards GMOs and takes a far less precautionary approach to regulating GMOs than European countries (Montpetit 2005; Andrée 2006).

According to one commentator, 'the official view in government is that transgenic organisms are not really all that different from non-GM food and crops.' This view is seen as being based on a 'purely scientific assessment, backed by international expert consultations' and it is argued that it 'should set the context for any policies dealing with GMOs' (Andrée 2002).

In keeping with this general enabling approach to GMOs, Canada, the world leader in canola production, was the first country to commercialise HT varieties of this crop in 1996. By 2018, 95% of its canola carried this trait (ISAAA 2018) and, as would be expected, it is also a leader in research into improved varieties of canola. Some of these include glufosinate-ammonium herbicide tolerance (to help prevent resistance to glyphosate), male sterility and varieties producing long-chain omega-3 fatty acids (ISAAA 2017b).

By the year 2017, 95% of the country's soybeans were HT, almost 100% of its maize crop was either both, or either one of, *Bt* and HT, and 100% of its sugar beet was HT. Three new crops have recently been commercialised. The first was low-lignin alfalfa stacked with the HT trait, which was commercialised in 2016. The decreased levels of lignin make it more digestible to livestock (Guo *et al.* 2001) and allow farmers to delay harvest by up to 10 days in order to obtain greater yield without losing quality (ISAAA 2018). Although the initial lignin reduction was achieved some years ago, development of the commercial product, HarvXtra™, required more than 15 years of research and field trials (Barros *et al.* 2019). The research was done by members of a partnership formed in 2002 between Forage Genetics International, based in Idaho, USA, the Noble Foundation, based in Oklahoma, USA, and the US Forage Research Centre, together with scientists from the universities of Wisconsin, Minnesota and the University of California, Davis. Success was achieved when a critical gene in the lignin biosynthesis pathway was suppressed (Forage Genetics International 2019).

The second new crop is the Innate® potato developed by the JR Simplot company. It has decreased levels of reducing sugars, reduced acrylamide potential (by reducing asparagine) and black spot bruising tolerance (Halterman *et al.* 2016). These first-generation Innate® potatoes received regulatory approval in 2016, but have since been improved upon by the second generation. This generation is also protected against the late blight pathogen, the fungus-like oomycete *Phytophthora infestans* (ISAAA 2017b). Simplot estimates this trait can result in up to a 50% reduction in fungicide applications annually. Reduced asparagine means that accumulation levels of acrylamide can be reduced by up to 90% when potatoes are cooked at high temperatures (Business Wire 2016).

The third crop on the market is the Arctic Apple produced by the Canadian firm Okanagan Speciality Fruits Inc. Gene silencing was used to reduce the expression of the enzyme polyphenol oxidase (PPO) that prevents the fruit from browning. It was approved by the Canadian Food Inspection Agency in 2017 (Canadian Food Inspection Agency 2019).

In addition, GM salmon went on sale in 2017, having been approved by the Canadian Food Inspection Agency in May 2016. It took the US company, AquaBounty Technologies, 25 years to bring their

AquaBounty salmon from discovery to commercialisation. The fish, a variety of Atlantic salmon (*Salmo salar*), has been engineered to grow faster than its conventional counterpart, reaching marketable size in roughly half the time (Waltz 2017).

So why has Canada 'got it right'? In a review of the state of GM regulation in Canada, Smyth (2014) writes that the key observation to take away from his article is that 'the market is capable of managing issues without government regulation and is able to inform government as to when legislation is required.' This situation is the polar opposite of the situation in the EU, where governments are attempting to regulate the most insignificant aspects of the technology and have created a situation of complete paralysis.' In addition, the role of the Canadian government highlights 'the actions of a responsible approach to the regulation of GM crops'. Moreover, it is highly unlikely that the Canadian company, Okanagan Specialty Fruits, would have spent so many years developing the Arctic Apple if they were not confident that the regulatory system would, indeed, use a science-based assessment when this product was sent to them for approval.

There is certainly a lesson here for countries wishing to allow their farmers access to this type of technology and local companies to invest in the development of new GM crops.

Argentina

GM crops are regulated in Argentina under the general Law on Seeds and Phytogenetic Creations. Their aim is to promote the efficient production and marketing of crops by providing farmers with assurances as to the identity and quality of seeds that they acquire while protecting the intellectual property of phytogenetic innovations. It provides a definition of seeds that is broad enough to include transgenic crops, because it includes all vegetable matter susceptible to sowing or propagation. The law is intended to promote the development and production of modern biotechnology by granting tax incentives to qualifying research and production projects that meet safety and health standards (Rodriguez-Ferrand 2014).

Argentina was among the first countries to use GM crops to improve its agriculture. The country's competitive agriculture is key to the economy and Argentina has a history of being an early adopter of all types of new agricultural technologies. Their use of GM crops was enabled by the creation in 1991 of the National Commission on Agricultural Biotechnology (CONABIA). The country first started growing GM crops in 1996 with the introduction of soybeans tolerant to glyphosate (Trigo 2011). Since then, it has increased its production of GM crops to become the third largest grower of biotech crops in the world, after the USA and Brazil (ISAAA 2018).

Argentina has been growing 'Intacta' soybeans since the 2013–14 crop year. This variety contains the *Bt* gene, *cry1Ac*, which confers resistance to lepidopteran insects, as well as the *cp4esps* (*aroA:CP4*) gene for glyphosate resistance. It accounted for 100% of the soybean 1st crop and 20% of the country's soybean 2nd crop in 2018 (Brookes 2018). There are two crops of soybeans per season, the 2nd being planted after a wheat crop in the same season. The combination of HT technology and the no-till production it allows (see Chapter 6), shortens the production season (time needed to prepare soil, plant, grow and harvest), allowing many farmers to plant a 2nd soybean crop. Before HT soybeans were available, less than 5% of the Argentina crop was 2nd crop. GM technology has therefore facilitated the expansion of this 2nd crop area to the current 20% (Graham Brookes, pers. comm., 5 November 2019).

However, Argentina, like so many other agriculturally intensive countries, is facing economic hardships due to an increased risk of drought prompted by global warming. This country is the largest exporter of soybean oil and soybean meal and these two commodities, together with maize, accounted for 17.7% of Argentina's total exports in 2017. Importantly, the country will begin commercialisation of the first GM drought and salt-tolerant soybeans in 2019 (Cerier 2018). How were they developed?

This was done by Rachel Chan, a senior researcher at Argentina's National Scientific and Technical Research Council, and her team (González *et al.* 2019). They used a gene from the sunflower that codes for a transcription factor, HaHB4, which triggers dehydration tolerance in *Arabidopsis*. The seed is owned by biOceres, an Argentine company involved in agricultural development. The gene was introduced into soybeans in 2012 and field tests showed them to be as nutritious as conventionally grown soybeans. The soybeans are able to tolerate periods of low water intake, show no toxicity to animals or humans and to have no negative effect on the environment (Cerier 2018).

One of the things that Argentina has done right is recognising at government level that GM crops have the potential to increase agricultural output. This is important because food exports are key to Argentina's gross domestic product (GDP). Therefore, the uptake of GM crops has the potential to improve the economy. Indeed, support of GM crops is a 'state policy', meaning that it has not changed as different political parties come into power.

In March 2017, the agriculture minister, Dr Luis Miguel Etchevehere, said that the promotion of GM crops was designed to increase the 'leadership of our country in the development, regulation, and safe and intelligent use of agricultural biotechnology' (Cerier 2018). This indicates that a country such as Argentina can also 'get it right' by recognising that if it invests in the development of this technology, the country will also grow in scientific prestige and leadership. Thus, a country that 'buys in' to GM crops can benefit additionally in ways that could lead to increased investment, which could result in improvements in job availability. Argentina has seen the advantages of GM crops instead of focusing on potential disadvantages.

Another aspect of Argentina's 'getting it right' was that as early as 2012 it was implementing regulatory measures aimed at speeding up the approval of new GM crops. This resulted in the time for regulatory assessment dropping from 42 to 24 months. According to the US Department of Agriculture's 2016 study, the new measures 'accomplished the expected goal of reducing the approval times and proved to be successful in reducing bureaucracy' (Cerier 2018).

This enabling environment helped biOceres, mentioned above, to develop GM alfalfa, called HarvXtra® Alfalfa, which contains less lignin. The company is owned by some of the largest farm operators and agro-industrial companies in Argentina (www.bioceres.com.ar). In addition, the company has developed drought-tolerant HB4 wheat in a joint venture with the French breeding company, Florimond Desprez, which contributed the germplasm for Bioceres's HB4 technology. The CEO of Bioceres, Mr Federico Trucco, has stated that they hoped to supply this wheat variety to farmers during the 2019–20 season if the government approves the trait. In an interview with eFarmNewsAr he stated: 'I hope government authorities realize that HB4 is a milestone for the scientific sector in the country and for the food and agricultural chain' (Patiño 2018).

Therefore, yet again, Argentina 'got it right', as shown by the development of successful new crops that will have a positive effect not only on biOceres' growth, but also on Argentina's economy and scientific prestige.

Brazil

In September 2003, Brazil decided to allow farmers to grow GM soybeans for a 1-year period. This decision did not find favour with Greenpeace Brazil and the Brazilian Green party, who, together with some non-governmental groups, announced that they would seek to have this decision overturned in the courts. They felt let down by President Luiz Inacio Lula da Silva, whose Workers Party had resisted GM crops when it was in opposition. Now, however, the President had apparently given in to agricultural officials who wanted a 1-year emergency measure because farmers in Rio Grande do Sul state had been smuggling large quantities of GM soya seed across the Argentine border for planting in Brazil (Vidal and Chetwynd 2003).

In support of the government's action, Bob Callanan, head of the pro-GM American Soybean Association, said, 'We have long been frustrated by Brazil growing illegal GM seeds. This would be a step towards allowing Monsanto to collect the fees due to it and help to end the paper shuffle where EU countries bought Brazilian foods and pretended that it was not GM' (Vidal and Chetwynd 2003).

Far from limiting its planting of GM crops to a single year, Brazil's farmers have grown them continuously and, in 2018, they grew the second largest area of these crops (51.3 million hectares), with the USA growing the largest. This came about because, as will be seen below, the government realised the profitability of growing these crops.

Initially the soybean crop was either *Bt* or HT and, at the end of the 1-year trial period, the government, realising its profitability, allowed planting to continue. In 2013, the stacked trait of *Bt*/HT Intacta™ was introduced into 4% of the crop and this innovation was so effective that by 2018 it had grown to 52%. The cultivation of these GM soybeans resulted in additional production during the 2013–18 period of 15.7 million tonnes, which translates into income gains of US$6.1 million (Brookes 2018). By 2018, soybeans accounted for 67% of the land on which GM crops are grown, with an adoption rate of 96% (ISAAA 2018). This is due both to its profitability as well as market demand, particularly from China, which buys large amounts for both food and animal feed.

Other GM crops are also grown in Brazil. In 2018, maize accounted for 30% of such crops with an adoption rate of 89%. *Bt* cotton was only planted in 2% of the GM crop fields and *Bt* sugarcane was grown for the first time in that year, accounting for the final 1% (ISAAA 2018). With this high acceptance of GM crops, due to their profitability and export value, it is not surprising that in a survey of 1250 rural producers conducted by the Brazilian Biotechnology Information Council, 90% recognised the importance of the *Bt* and HT traits (ISAAA 2017a).

The development of the new *Bt* sugarcane, resistant to its main insect pest, the borer *Diatrarea saccharalis,* was achieved by a local biotechnology company, Centro de Tecnologia Canavieira (CTC). The event, CTC20BT, has been approved by the Brazilian national technical biosafety commission, CTNBio. The variety contains the same *Bt* gene widely used for more than 20 years in soybean, maize and cotton. CTC has conducted tests on the products of CTC20BT and found that both the sugar and ethanol obtained from it are identical to that derived from conventional sugar. Environmental studies have also revealed no adverse effects on soil composition, sugarcane biodegradability or insects other than the target pests (ISAAA 2017a).

There is no doubt that the reason for CTC going ahead to develop this *Bt* sugarcane was the knowledge that the Brazilian government, as with the Argentinian government, supports the development of GM crops. However, this has taken some time to develop. Until 2005, Brazil's regulatory system, both for research and commercial release, was somewhat confusing. In 1998, the commercial approval of Roundup Ready® soybeans by CTNBio resulted in 6 years of largely non-scientific debate. Fortunately, a new law came into being in 2005, which allowed for a much more streamlined CTNBio. This enabled the first plantings of transgenic soybeans to be harvested. The new law was a great improvement and, apart from the establishment of safety standards linked to enforcement measures, it fosters scientific advances in the areas of biosafety and biotechnology. The law also created the National Biosafety Council, which analyses the socioeconomic effects of GMOs and may revoke a CTNBio decision to commercially release a new biotechnology product if its risk assessment reveals problems (De Souza *et al.* 2013).

Thus, Brazil 'got it right' by first allowing a 1-year trial planting period, then agreeing to its continuation and, in 2005, putting in place a regulatory system that was sufficiently strong to enable farmers to plant GM maize, cotton and sugarcane commercially.

To conclude this section on the countries that 'got it right', what do they have in common? They all have governments that are sympathetic to the needs of their farmers in countries where agriculture forms a major part of their GDP. They have all developed flexible regulatory systems that work on a case-by-case basis and whose decisions are science-based. As a result, the regulations do not act as barriers to the development and implementation of GM crops. In some cases, their laws are intended to promote the development and production of modern biotechnology by granting tax incentives to qualifying research and production projects that meet safety and health standards. In addition, in many cases, these enabling environments have sparked the development of strong private innovation sectors, which work with publicly funded institutions to develop GM crops specific to the needs of their farmers. Another factor that they all have in common is the importance of exports of GM crops, which results in a significant improvement of their GDPs. Whether public opinion of GM crops is also related to this factor is a question that still has to be answered.

Now let us look at two countries that tried and nearly got it right: China and Burkina Faso.

China

Since 1997, China has approved 64 GM crop events, including canola, cotton, maize, papaya, petunia, poplar, rice, soybeans, sugar beets, sweet pepper and tomatoes (ISAAA 2017b). This included phytase maize that has increased phosphorus levels, which is good for livestock (Shen *et al.* 2008; Xu *et al.* 2018). That sounds impressive, so why, then, is China not among the top GM crop producers in 2019? This story is told in a fascinating book by Cong Cao (2018) and reviewed comprehensively by Ron Herring (2019).

China began growing *Bt* cotton in the 1990s, relying on its public sector for its development. There was rapid uptake of the technology in those early years and, indeed, China was the first country outside the West to build significant capacity in biotechnology in its state-led institutions. The government stated that it wished 'to become a global leader in biotechnology' (Cao 2018). Its priority in agricultural biotechnology was to reduce the widespread pesticide pollution, improve farmers' incomes and decrease the country's dependence on food importation by increasing local production. This aim was achieved to a large extent with *Bt* cotton, as conventional cotton is notorious for its heavy use of insecticides, but what about food crops?

Before we get to that, let us look briefly at a problem found in many developing countries with respect to GM crops – that of illegal trafficking in GM seeds (mentioned above for Brazil and which will be covered more extensively in Chapter 10). Chinese farmers were the first in the world, in 1988, to grow virus-resistant tobacco, but many of the seeds were illegally obtained. This problem is not uncommon, as is explained by Ron Herring (2019), when the benefits of GM crops to farmers drive them to illegal actions where the official regulations or private sector financial claims are onerous. 'There is a soft global law – the Cartagena Protocol – on international trade, but it has no eyes or ears in the villages, nor enforcement machinery; national governments have few incentives to report their failures, nor any real political interest in ferreting out illicit seeds … It is not easy to claw back a useful technology from farmers once they have used it' (Herring 2019). Thus, quite a large sector of the Chinese farming community was growing GM crops illegally. Although this might help farmers, it places the regulatory authorities in a difficult position because it shows they are preventing their own people from benefiting from new technology. In the case of Brazil, where illegal GM soybean seeds were being obtained from Argentina, the government decided to allow them to be grown. As a result, Brazil is second only to the USA in its uptake of GM crops. The opposite seems to have happened in China. The government appears to have taken a decision not to allow farmers to plant GM seeds. One wonders what will happen in the future to farmers planting illegal seeds.

Now to return to the question of why China did not pursue GM crops for food production. The answer can be answered briefly by the term 'risk aversion'. As in many countries in the 2000s, Greenpeace became a major player in China and the spectre of risk was high on their agenda. Social media began to echo their scare stories and soon the government was caught between the desire to see the spread of agricultural biotechnology and apparent consumer scepticism. This was fuelled by fringe scientists giving credence to the alarming stories being circulated. And the anti-GM crop stance taken by Europe had its effect in China as well. The potential of future risks and unknown consequences of planting GM crops played into the hands of the anti-GMO opposition and thus China, after its initial early entry into GM crops, has dropped by the wayside (Cao 2018; Herring 2019).

Burkina Faso

In 2008 farmers in Burkina Faso, one of Africa's largest cotton producers, began to plant *Bt* cotton commercially. By 2014, ~74% of cotton grown in the country was *Bt* and grown by an estimated 140 000 smallholder farmers. Advantages included a 20% yield increase compared with conventional cotton, a reduction in insecticide use of ~67% and an estimated profit increase of US$64 per hectare, despite the high cost of the seed (James 2014).

However, problems began to appear. The cotton produced by the *Bt* varieties had shorter staples (fibres), producing less desirable lint and thus resulting in an inferior quality product. Although this was not a problem for the farmers, it was for the cotton ginning companies. The combination of shorter staples and lower lint quality undermined their profits. As a result, the cotton companies, which also control the provision of seeds to farmers, unilaterally phased out *Bt* cotton, much to the dismay of many farmers (The Conversation 2016).

Was this a good move? In 2019 the Interprofessional Cotton Association of Burkina (AICB), an industry body comprising farmers and other sector players, reviewed the situation. They had set a production target of 800 000 tonnes for the 2018–19 cotton season, but the country produced just 436 000 tonnes. This was in spite of the AICB offering farmers a record US$27.4 million in incentives in the form of subsidies on insecticides, fertilisers and irrigation facilities. The country, which was formerly Africa's largest cotton producer, was, in 2019, fourth, behind Côte d'Ivoire, Mali and Benin (Gakpo 2019).

The dire position of the Burkina Faso cotton industry was presented at an AICB media briefing in the capital, Ouagadougou. Among the factors put forward to explain the decline in production were regional farmer boycotts over unfair treatment, insecurity resulting from terrorist attacks and bad weather. However, many also blamed the return to conventional seeds as a result of the phasing out of *Bt* cotton, resulting in increased pest attacks.

The government has announced new measures to improve the cotton industry, which include farmer subsidies and increases in the price that processors pay for cotton from farmers, as well as cheaper fertilisers and insecticides. However, scientists say that the industry will continue to perform poorly if *Bt* cotton is not reintroduced. The National Union of Cotton Producers of Burkina Faso (UNPCB) issued a press statement in February 2018 in which it stated that, 'UNPCB is for genetically modified cotton because we are all aware of its benefits … UNPCB is committed to finding a solution quickly for the return of the GM cotton together with the cotton companies and the government' (Gakpo 2019).

Whether the government's plans of incentivisation etc. will prove successful or whether the cotton industry can only be revived by a return to *Bt* cotton remains to be seen. Will Burkina Faso be an example of a country that didn't need to 'get it right' or will it be one that does, in the end, 'get it right? Time will tell.

Another issue that in some countries is hindering the uptake of GM crops is that of labelling. The next chapter will consider this thorny issue.

References

Andrée P (2002) The biopolitics of genetically modified organisms in Canada. *Journal of Canadian Studies. Revue d'Etudes Canadiennes* **37**, 162–191. doi:10.3138/jcs.37.3.162

Andrée P (2006) An analysis of efforts to improve genetically modified food regulation in Canada. *Science & Public Policy* **33**, 377–389. doi:10.3152/147154306781778885

Barros J, Temple S, Dixon RA (2019) Development and commercialization of reduced lignin alfalfa. *Current Opinion in Biotechnology* **56**, 48–54. doi:10.1016/j.copbio.2018.09.003

Brookes G (2018) The farm level economic and environmental contribution of Intacta soybeans in South America: the first five years. *GM Crops & Food* **9**, 140–151. doi:10.1080/21645698.2018.1479560

Business Wire (2016) 'Innate® second generation potato receives FDA safety clearance'. <https://www.businesswire.com/news/home/20160113005894/en/Innate%C2%AE-Generation-Potato-Receives-FDA-Safety-Clearance>

Canadian Food Inspection Agency (2019) 'Questions and answers: Arctic Apple'. <http://www.inspection.gc.ca/plants/plants-with-novel-traits/general-public/questions-and-answers-arctic-apple/eng/1426884802194/1426884861294>

Cao C (2018) *GMO China: How Global Debates Transformed China's Agricultural Biotechnology Policies.* Contemporary Asia in the World. Columbia University Press, New York, NY.

Cerier S (2018) 'Argentina and GMOs: Exploring the nation's long relationship with biotech crops'. Genetic Literacy Project, 6 September 2018. <https://geneticliteracyproject.org/2018/09/06/argentina-and-gmos-exploring-the-nations-long-relationship-with-biotech-crops/>

De Souza GD, de Melo MA, Kido EA, de Andrade PP (2013) The Brazilian GMO regulatory scenario and the adoption of agricultural biotechnology.

The World of Food Science. <https://worldfoodscience.com/article/brazilian-gmo-regulatory-scenario-and-adoption-agricultural-biotechnology>

Forage Genetics International (2019) 'Get xtra control and xtra quality where it counts'. <https://www.foragegenetics.com/Products-Technologies/HarvXtra-Alfalfa>

Gakpo JO (2019) 'Burkina Faso cotton production plummets after phasing out GMO crop'. Alliance for Science, 26 April 2019. <https://allianceforscience.cornell.edu/blog/2019/04/burkina-faso-cotton-production-plummets-phasing-gmo-crop/>

Gastrow M, Roberts B, Reddy V, Ismail S (2018) Public perceptions of biotechnology in South Africa. *South African Journal of Science* 114, 2017-0276. doi:10.17159/sajs.2018/20170276

González FG, Capella M, Ribichich KF, Curín F, Giacomelli JI, Ayala F, Watson G, Otegui ME, Chan RL (2019) Field-grown transgenic wheat expressing the sunflower gene *HaHB4* significantly outyields the wild type. *Journal of Experimental Botany* 70, 1669–1681. doi:10.1093/jxb/erz037

Gouse M, Sengupta D, Zambrano P, Zapeda JF (2016) Genetically modified maize: less drudgery for her, more maize for him? Evidence from smallholder maize farmers in South Africa. *World Development* 83, 27–38. doi:10.1016/j.worlddev.2016.03.008

Green JM (2018) The rise and future of glyphosate and glyphosate-resistant crops. *Pest Management Science* 74, 1035–1039. doi:10.1002/ps.4462

Guo D, Chen F, Wheeler J, Winder J, Selman S, Peterson M (2001) Improvement of in-rumen digestibility of alfalfa forage by genetic manipulation of lignin O-methyltransferases. *Transgenic Research* 10, 457–464. doi:10.1023/A:1012278106147

Halterman D, Guenthner J, Collinge S, Butler N, Douches D (2016) Biotech potatoes in the 21st century: 20 years since the first biotech potato. *American Journal of Potato Research* 93, 1–20. doi:10.1007/s12230-015-9485-1

Herring RJ (2019) [Review] Cao Cong, *GMO China: How Global Debates Transformed China's Agricultural Biotechnology Policies.*' H-Asia H-Net Reviews in the Humanities and Social Sciences, September 2019. <http://www.h-net.org/reviews/showrev.php?id=53363>

International Survey of Herbicide Resistant Weeds (2019) List of herbicide resistant weeds by country. <http://www.weedscience.org/Summary/Country.aspx?CountryID=37>

ISAAA (2017a) 'Brazil approves GM sugarcane for commercial use'. Crop Biotech Update, 14 June. <http://www.isaaa.org/kc/cropbiotechupdate/newsletter/default.asp?Date=6/14/2017>

ISAAA (2017b) 'Global status of commercialized biotech/GM crops in 2017: biotech crop adoption surges as economic benefits accumulate in 22 years'. ISAAA Brief No. 53. ISAAA, Ithaca, NY.

ISAAA (2018) 'Global status of commercialized biotech/GM crops in 2018: biotech crops continue to help meet the challenges of increased population and climate change'. ISAAA Brief No. 54. ISAAA, Ithaca, NY.

James C (2014) 'Global status of commercialized biotech/GM crops: 2014'. ISAAA Brief No. 49. ISAAA, Ithaca, NY.

Kotey DA, Assefa Y, Obi A, van den Berg J (2016) Disseminating genetically modified (GM) maize technology to smallholder farmers in the Eastern Cape province of South Africa: extension personnel's awareness of stewardship requirements and dissemination practices. *South African Journal of Agricultural Extension* **44**, 59–74. doi:10.17159/2413-3221/2016/v44n1a370

Law Library of Congress (2012) *Restrictions on Genetically Modified Organisms: Canada*. <https://www.loc.gov/law/help/restrictions-on-gmos/canada.php>

Montpetit E (2005) A policy network explanation of biotechnology policy differences between the United States and Canada. *Journal of Public Policy* **25**, 339–366. doi:10.1017/S0143814X05000358

Patiño JP (2018) 'Will Argentina be the first country approving a GMO wheat?' eFarmNew*sar*, 16 November 2018. <https://efarmnewsar.com/2018-11-16/will-argentina-be-the-first-country-approving-a-gmo-wheat.html>

Shen Y, Wang H, Pan G (2008) Improving inorganic phosphorus content in maize seeds by introduction of phytase gene. *Biotechnology (Faisalabad)* **7**, 323–327. doi:10.3923/biotech.2008.323.327

Smyth SJ (2014) The state of genetically modified crop regulation in Canada. *GM Crops and Food: Biotechnology in Agriculture and the Food Chain* **5**, 195–203. doi:10.4161/21645698.2014.947843

The Conversation (2016) 'Lessons to be learnt from Burkina Faso's decision to drop GM cotton'. The Conversation Australia, 5 February 2016. <https://theconversation.com/lessons-to-be-learnt-from-burkina-fasos-decision-to-drop-gm-cotton-53906>

Thomson JA (2013) *Food for Africa: The Life and Work of a Scientist in GM Crops*. UCT Press, Cape Town, South Africa

Trigo EJ (2011) 'Fifteen years of genetically modified crops in Argentine agriculture'. Consejo Argentino para la Información y el Desarrollo de la Biotecnología, November 2011. <http://www.argenbio.org/adc/uploads/15_years_>

Vidal J, Chetwynd G (2003) 'Brazil agrees to grow GM seed'. *The Guardian*, 26 September 2003. <https://www.theguardian.com/science/2003/sep/26/gm.food>

Waltz E (2017) 'First genetically engineered salmon sold in Canada'. *Scientific American*, 7 August 2017. <https://www.scientificamerican.com/article/first-genetically-engineered-salmon-sold-in-canada/>

Xu XH, Guo Y, Sun H, Li F, Yang S, Gao R, Lu X (2018) Effects of phytase transgenic maize on the physiological and biochemical responses and the gut microflora functional diversity of *Ostrinia furnacalis*. *Scientific Reports* **8**, 4413. doi:10.1038/s41598-018-22223-x

8

To label or not to label, that is the question

Why *is* there a question? Surely consumers have an intrinsic right to information about the products on supermarket shelves? Indeed, this is not disputed if the information is simply about the contents of the item. But if it covers the way in which that item was produced, value judgements come into play. Value system-based labelling of food products is, of course, nothing new. Take, for instance, religion-based labels such as 'halal', 'kosher' or ethics-based labels such as 'free range' or 'organic'; there are very few questions about these. However, these are all voluntary labels managed and maintained by the relevant interest groups. They are not mandatory legislated labels affecting all customers, especially in their pockets (Biosafety SA 2019). Therefore, another question that needs to be answered is, 'Who will pay for the labels?'

People might also ask, 'Why not just place a GM label on all food products and be done with it?' One of the reasons is that the wording on the label can prejudice the consumer against that product, leading to unfair discrimination. Therefore, questions that need to be answered are, 'Is the label useful?' 'Is the label educational?' Or 'Is the label simply malicious and confusing?' What about a consumer who knows little about the GMO debate and reads a label that has the words 'Made from genetically modified maize' or 'Grown from seed developed through genetic engineering'? Such labels could be extremely confusing and the purchaser might conclude that if it's labelled in this way it must be dangerous.

In addition, a further question arises: 'What effects will labelling in the West have on developing countries?' We will consider all these issues in this chapter.

Why is food labelled?

Food is labelled in order to provide consumers with important information about the attributes of the product to enable them to make informed decisions. At least, that is the theory. Problems arise if consumers do not understand what the label means. How many potential buyers know the difference between saturated fats, trans-fats or monounsaturated and polyunsaturated fats? Most of them probably know that saturated fats are 'bad' and base their choices on that. But when it comes to the process whereby an ingredient in the food item was derived, how many understand that? Of course, if the label said: 'Made using child labour', 'Contains products derived by chemical mutagenesis' or 'May still carry insecticide residues' (on a bag of apples), most consumers would readily understand. However, those situations do not require a label.

But if the label reads 'Contains GMOs', how many consumers would really understand what this means? How many will know that, despite the label, the food in question has been declared 'substantially equivalent' to the same food made by conventional breeding? In 1993 the OECD produced a report entitled 'Safety evaluation of food derived by modern biotechnology' (OECD 1993). They concluded: 'if a new food or food component is found to be substantially equivalent to an existing food or food component, it can be treated in the same manner with respect to safety. No additional safety concerns would be expected.' Many countries have accepted this guideline to distinguish between novel foods and their traditional counterparts (Schauzu 2000).

The simple answer to the question 'Why is food labelled?' should then be that if the food is substantially equivalent it need not be labelled. But things in the GM world are seldom simple and substantial equivalence is just one example. Critics of GM crops, as discussed by Brazeau (2019), often think that this term must be some sort of legalistic loophole. As he writes, 'They ask how something can be so novel that on one hand, it merits the legal protection of patent monopoly, and then on the other hand, the FDA can declare it to be substantially equivalent to its parent variety or breed.'

This conundrum can be cleared up if we consider another misconception. The scientific consensus is not about GMOs or GM crops, but about genetic engineering (GE). The consensus is about the *process*, not the *products*. In addition, what many people do not understand is that to a scientist there is no such thing as absolute safety, there is only relative risk. To scientists, GE is no more risky than traditional breeding methods, although most of them view it as more precise and therefore carrying a lower risk of harmful unintended consequences. In fact, because of the extreme regulatory scrutiny that GMOs and GM crops receive, they come with fewer risks for consumers, although, in either case, the risks are miniscule (Brazeau 2019).

Two examples of unintended consequences as a result of traditional breeding are celery and potatoes in the 1960s. Celery breeders developed a variety that produced higher levels of a natural insecticide called psoralen. However, when farm workers developed rashes due to dermatitis, the product had to be withdrawn. Potato breeders developed a variety that produced higher levels of another insecticide, solanine, but that turned out to be toxic to humans, so it was also withdrawn (Brazeau 2019).

Of course, some GM crops are intentionally designed not to be substantially equivalent. If Golden Rice with its increased levels of β-carotene, resulting in increased levels of vitamin A in consumers, were to be marketed in the USA, it would definitely need to be labelled. In this case, however, the producers will be only too delighted to label their product because it would help to alleviate health problems such as blindness and infertility. This will be covered in greater detail in Chapter 10.

The debate over whether foods derived from GM crops that are sold commercially, whether on supermarket shelves or elsewhere, should be labelled has not stopped. The public usually refers to such food as GM food, but strictly speaking, the only GM foods that have been on the market for several years are maize and soybeans. More recently, Arctic® apples, which do not brown when cut, or Innate potatoes, which produce lower levels of harmful acrylamide during frying, have become available in parts of the USA and Canada. To call these products GM food would be correct. However, most items that would be considered for labelling are those that might contain ingredients derived from a GM crop.

We often hear: 'It is my right to know what is in the food I am buying.' Or: 'If this is so safe why can't it be labelled?' These questions sound so simple and surely the answer is 'Let's label'? Unfortunately, although the questions are simple, the issues are not. For instance, as mentioned above, if the product is labelled on the basis of how it is made, shouldn't foods that have been heavily sprayed with insecticides and probably still carry toxic residues, also be labelled? Why stop with labelling produce made by genetic modification? Unfortunately, arguments such as these fall on deaf ears and, as will be seen below, several countries require mandatory labels for GMOs, while others allow voluntary labels.

One of the problems raised by such policies is that mandatory labelling of a product that is not dangerous is not part of a government's mandate. If products have passed the approval process that declares them as safe, why should the government require them to be labelled? In fact, it could be argued that this weakens the informative power of food safety labels. In addition, it signals to consumers that GMOs, in themselves, are a food safety risk (Lusk *et al.* 2018; McCluskey and Wesseler 2018; Zilberman *et al.* 2018).

Some countries set minimum standards for the amount of GM items in a given product. What, then, should the limit be of this amount allowed by the regulatory authorities for either mandatory or voluntary 'non-GM' labelling, and on what will that be based? In the case of contaminants, certain levels are allowed in products. Some examples, according to the US Food and Drug Administration (FDA 2018a), are frozen broccoli ($\geq$ 60 aphids per 100 g), cornmeal ($\geq$ 1 whole insects or equivalent per 50 g), macaroni or noodles ($\geq$ 225 insect fragments per 225 g in $\geq$ 6 subsamples) or canned peaches ($\geq$ 3% fruit by count is wormy or mouldy). Or, according to the Food and Agricultural Organization/World Health Organization's committee on contaminants in foods, 1 mg of arsenic can be present in 1 kg of edible foods and oils or 15 µg of aflatoxins (produced by fungal infection) in 1 kg of peanuts intended for further processing or 0.5 mg of melamine in 1 kg of liquid infant formula are allowed (FAO/WHO 2011). How would the label handle the adventitious presence of a GM item in a food product? Would it be done by allowing a certain percentage as a 'contaminant'? And if so, would this not be seen by producers (and members of the public who understand these regulations) as 'demonising' the technology?

What is to be done about food that is bought and consumed in developing countries? Much of it is not packaged and therefore cannot be labelled. In addition, there is a certain amount of illiteracy in some countries, so what is to be done about people who cannot read? And what about the people who do not understand the label even if they can read it? For instance, in South Africa only ~14% of the population in a recent poll could explain what 'GM' or 'GM food' means (Gastrow *et al.* 2018).

And what about truth in advertising? I have seen a 'GM-free' label on a shelf of brazil, cashew and hazelnuts at a wholesale market, despite the fact that there are no GM versions of these nuts. Does this amount to unfair advertising? A friend has even told me that she saw 'GM-free condoms' for sale! And what about canola oil, which is highly processed and no longer contains any DNA? It certainly does not contain any GMOs, alive or dead. Similarly, what about sugar derived from insect-resistant sugarcane from Brazil, as discussed in Chapter 7?

Although these examples are factually correct, even if there is no GMO equivalent of the item, it is clearly a marketing ploy. In the USA consumers often encounter a Non-GMO Project label. Companies that become members of this project pay an annual fee for each item displaying the label, which could even include minerals such as salt, or cat litter (Goldy 2019). As quoted by McCluskey and Wesseler (2018) in the introduction to a special issue of the journal *Food Policy*, the then US Agricultural Secretary stated, 'I knew it was all about marketing when I went to the grocery store and there was non-GMO, organic shampoo.'

Before we can answer the above questions, let us look at how labels are paid for.

Who pays for the label?

The answer to the question of who will pay is, of course, the consumer. The cost of simply putting a label on a product might be negligible, but the cost of complying with requirements of the preservation of the identity of a non-GM food and its subsequent certification will considerably increase the cost of the item. In addition, there seem to be almost irreconcilable differences in perceptions as to the effects that GM food can have on health and the environment. For instance, the US National Academies Committee on Genetically Engineered Crops reports that there is no substantiated evidence for a difference in risks to human health between currently commercialised GE crops and conventionally bred crops. In addition, they did not find conclusive cause-and-effect evidence of environmental problems from GE crops (National Academies of Sciences, Engineering and Medicine 2016). However, this evidence appears to be ignored by many consumers who seem to believe that non-GM food is of a higher quality and safer to eat. As discussed by Tosun and Schaub (2017), labelling has become a consumer issue rather than a human and/or environmental safety issue.

According to basic economic theory, mandatory labelling will result in increased food prices, because, in addition to the physical labelling, product identity preservation steps, which include monitoring and enforcement, will result in price increases. Therefore, a voluntary non-GMO labelling policy could be the preferred option both from an economic and food safety perspective (McCluskey and Wesseler 2018). This would certainly be true in developing countries where food security is often out of reach for many of their poorer inhabitants. It is all very well for wealthy countries, such as those in the EU, to demand labelling, but as has been discussed in Chapter 3, what the EU does has a major effect on what countries in the developing world do.

Let us now consider what some countries are doing about labels.

What countries are doing about labelling

In looking at what some of the countries growing the largest amounts of GM crops are doing about labelling, I have included Western countries whose actions could influence decision makers in developing countries.

The European Union

In the EU, all food (including processed food) or feed that contains more than 0.9% of the approved content being derived from a GM plant must be labelled. Due to concerns regarding the coexistence of GM and non-GM crops, all GM food must be traceable to its origin. Products of animals fed GM feed are not required to be labelled, which includes milk, meat and eggs. Labelling is required for vegetable oils and other highly refined products, even where the GM DNA or the proteins derived from such DNA are no longer present or detectable in the final product. The adventitious presence of any GM ingredient at levels below 0.9% does not require labelling (European Commission 2018).

The USA

In 2018, the US Department of Agriculture issued its first compulsory, national disclosure requirements for foods that have been altered in a way that cannot occur naturally. The guidelines, which use the term 'bioengineered' instead of the more commonly used 'genetically modified,' allow disclosure of bioengineered ingredients in several formats: in text, a symbol, a digital link printed on the packaging or text message that includes a statement on the package that instructs consumers how to receive such a message (USDA 2018). For example, companies can use a code with a statement such as: 'Scan here for more food information'. Customers who scan will reach a website where GM foods are disclosed. A diagram, such as shown in Fig. 8.1, can be used in black and white or in colour.

Food items that are derived from GM plants must meet the same safety, labelling and other regulatory requirements that apply to all foods regulated by the FDA. The agency has also noted that, since 1992, it has not become aware of any information showing that foods derived from GM plants differ in any meaningful way from other foods. 'These foods also don't present different or greater safety concerns than their non-genetically engineered counterparts. However, if a food derived from a genetically engineered plant is materially different from its traditional counterpart, the labelling of that food must disclose such differences' (FDA 2018b).

The rule exempts highly refined items, such as sugar or oil derived from a GM crop, because they do not contain detectable amounts of the modified genes. Thus, cold drinks and sweets will not carry a label.

Fig. 8.1. A disclosure diagram as suggested by the United States Department of Agriculture (USDA 2018). Figure used with permission from USDA Food Safety and Inspection Service.

The list of ingredients that manufacturers must currently disclose, unless they can demonstrate that they have not been produced from GM plants, comprises alfalfa, apples, canola, maize, cotton, eggplant, papaya, potatoes, soybeans, sugar beets and summer squash. The date for compliance was 1 January 2020, with an extension for small food manufacturers until 1 January 2021, although the mandatory date is 1 January 2022. Thus, the Non-GMO Project, mentioned above, should have ceased on 31 December 2019.

Also, in 2018, the USFDA issued guidance for manufacturers that wish to voluntarily label their foods as not containing ingredients derived from GE plants. They recommended such manufacturers use terms such as 'not genetically engineered', 'not bioengineered' or 'not genetically modified through the use of modern biotechnology'.

Considering the effect that the issue of whether or not to label has had on sentiment towards GM crops in the USA, Mark Lynas (2018) recalls that the anti-GMO movement in the USA really got going only when activists discovered the power of the labelling demand, both as a political organising tool and as a way of heightening fear among consumers: 'Why would they refuse to label GMOs in your food unless they knew they were dangerous?' Now that labelling has become compulsory in the USA, it will be interesting to observe what effect this has on pro- or anti-GMO attitudes in that country.

Canada

In Canada, products derived from GM crops are regulated as part of its existing framework for 'novel' products. Thus, their focus is on the traits in the products and not on the method whereby these traits were introduced. The Canadian Food Inspection Agency is responsible for regulating GM plants and for approving GM animal feed. Health Canada is responsible for assessing the safety of foods for human consumption. Indicating the presence or absence of products derived from GM crops on the label is voluntary unless there is a health or safety concern. These rules apply to all novel foods, whether derived from GM crops or not (Library of Congress 2014).

Argentina

Although there have been attempts by Argentina's congress to pass labelling legislation, there is currently no legally mandated label for GM food. The Ministry of Agriculture says this is because 'any food product obtained through biotechnology and is substantially equivalent to a conventional food product should not be subject to any specific mandatory label' (Cerier 2018).

Brazil

Mark Lynas has written an interesting piece on labelling of GM products in Brazil (Lynas 2018). In it he highlights the label that, according to Brazilian law since 2003, has been required on all consumer products containing GM material. The design of the label, a black T-shape within a bright yellow triangle, is similar to the common biohazard warning signs, suggesting that GMOs are something to be feared.

In an article entitled '15 years of genetically modified organisms (GMO) in Brazil: risks, labelling and public opinion' (Castro 2016), the author concludes that 'the ability of the consumer to decide whether to accept GMOs is merely a discourse of deterrence, which seeks to hide the fact that the State has already authorized its cultivation'. The article also notes that soybean processing companies that certify their products as GM-free had failed to receive any additional gain in the prices of their products. Perhaps this is due to the fact that the survey of over 800 consumers revealed that they were far more concerned with issues related to biological and chemical contamination, as well as the nutritional characteristics of the food they consumed, than any issues related to plant biotechnology.

South Africa

In 2004 the South African Department of Health (DOH) introduced labelling regulations whereby food produced through genetic modification that differed from the equivalent food produced by conventional methods had to be labelled. The differences could relate to composition, nutritional value, mode of storage, preparation or cooking, allergenicity or genes with animal or human origin. However, as all the current GM foods are considered to be substantially equivalent, these regulations have never been triggered (Gouws and Groenewald 2015).

In contrast, in 2008 a bill was passed by the Department of Trade, Industry and Competition (dtic) that required that all foods derived from GM crops be labelled. The difference between the DOH and dtic regulations is that the former is based on health and food safety issues while the latter hinges on the consumer's right to information (Gouws and Groenewald 2015).

Food producers, importers and packagers have to choose one of three mandatory labels: (i) 'contains genetically modified ingredients or components' if the GM content is at least 5%, (ii) 'produced using genetic modification' for food that has been produced directly from GMO sources or (iii) 'may contain GMOs' when it is scientifically impractical and not feasible to test food for GM content. There are also voluntary labels, which include: (i) 'does not contain GMOs' where the GM content is less than 1%, (ii) 'GM content is less than 5%' where the GM content is between 1% and 5% or (iii) 'may contain genetically modified ingredients' if it can't be detected (Biosafety SA 2019).

The Gouws and Groenewald (2015) article goes on to question whether the cost that these regulations impose, which will be covered by the consumer, and the possibility of unfair discrimination, are warranted. For example, between 80% and 90% of South Africa's locally produced maize is GM. The authors calculate that the average cost of labelling, which requires the testing for the presence of any GM products and their separation along the value chain, will be between 9% and 12% of the commodity (Biosafety SA 2019). Is this fair?

In addition, industry fears that these labels will be used to promote unfair discrimination against the developers, growers and marketers of

GM crops under the guise of 'consumer choice'. The article goes on: 'GM technology and, in particular, GM-derived foods have long been the target of destructive campaigns organised by NGOs and individuals with self-declared "environmental" and/or "social" agendas.' They base this on the fact that GM technology is strictly regulated by the GMO Act of 1997, which requires all GM products to pass a rigorous biosafety risk assessment before commercialisation.

Concluding this section on countries' labelling practices, all except Argentina and Canada, require labelling. The EU is the most stringent, requiring labels even on products such as oils and other refined products where no residual DNA or proteins derived from a GM crops can be detected. That these approaches will have an effect on decision makers in the developing world is likely, bearing in mind the influence that the EU has, particularly on countries in Africa, as discussed in Chapter 3.

Finally, a recent study looked at the effects of GMO labelling in the USA, where mandatory GM labelling legislation has recently come into effect (Yeh *et al.* 2019). The authors point out that food labels can be more than just an identifier of a product's attributes; they are also a signal that can influence consumer preferences for alternative products available in the market. One major finding was that 'when GM-labelling is mandatory, non-GM products may not need the not-GM label because consumers will perceive the non-labelled products as not containing GMs, which will of course be true.' As the study was concerned with only fresh produce, the author states, 'GM fresh produce growers must ensure that the positive attributes brought by GM technology (e.g. avoiding rapid browning in apples) off-set the negative effect on demand caused by the mandatory GM label.'

Can we now answer the question posed at the beginning of this chapter: 'What effects will labelling in the West have on developing countries?' The fact that the EU not only requires labels on all food that contains more than 0.9% GM origin, but also demands that all GM-containing food be traceable to its origin, must certainly indicate to people in the developing world that this food is, at the very least, questionable. This would be reinforced by the now mandatory labelling in the USA, despite the fact that the FDA acknowledges that the 'substantial equivalence of all crops, regardless of how they were bred, is evaluated using the same standards' and that 'GMO crops currently on the market are as safe as their non-GM counterparts' (Genetic Literacy Project 2018).

In conclusion, as more and more countries demand that GM products be labelled, the benefit–risk ratios will become increasingly important. Developers of such products need to be mindful of this and use these data more convincingly in the way in which they develop their markets.

In the next chapter, the influence of Western countries on Africa will be discussed. Why is it that only four countries on this continent are growing GM crops commercially? And can we learn anything from those that are currently at the stage of field trials but are not prepared to take the next step?

References

Biosafety SA (2019) 'The labelling of GM foods in South Africa'. <http://biosafety.org.za/information/know-the-basics/gmo-safety/the-labelling-of-gm-foods-in-south-africa>

Brazeau M (2019) 'GMOs are 'substantially equivalent' to conventional foods. Should they face reduced regulations?' Genetic Literacy Project, 22 November 2019. <https://geneticliteracyproject.org/2019/11/22/gmos-substantially-equivalent-to-conventional-foods-should-reduced-regulations/?mc_cid=c0dd67a18b&mc_eid=832c481596>

Castro B de S (2016) 15 years of genetically modified organisms (GMO) in Brazil: risks, labelling and public opinion. *Agroalimentaria* **42**, 103–117.

Cerier S (2018) 'Argentina and GMOs: exploring the nation's long relationship with biotech crops'. Genetic Literacy Project, 6 September 2018. <https://geneticliteracyproject.org/2018/09/06/argentina-and-gmos-exploring-the-nations-long-relationship-with-biotech-crops/>

European Commission (2018) *GMO legislation*. <https://ec.europa.eu/food/plant/gmo/legislation_en>

FAO/WHO (2011) *Food Standards Programme Codex Committee on Contaminants in Foods*. The Hague, The Netherlands. <http://www.fao.org/tempref/codex/Meetings/CCCF/CCCF5/cf05_INF.pdf>

FDA (2018a) *Food Defect Levels Handbook*. US Food and Drug Administration, 7 September 2018. <https://www.fda.gov/food/ingredients-additives-gras-packaging-guidance-documents-regulatory-information/food-defect-levels-handbook>

FDA (2018b) Guidance for industry: guide for developing and using data bases for nutrition labelling. US Food and Drug Administration, 20 September 2018. <https://www.fda.gov/regulatory-information/search-fda-guidance-documents/guidance-industry-guide-developing-and-using-data-bases-nutrition-labeling>

Gastrow M, Roberts B, Reddy V, Ismail S (2018) Public perceptions of biotechnology in South Africa. *South African Journal of Science* **114**, 2017-0276. doi:10.17159/sajs.2018/20170276

Genetic Literacy Project (2018) 'Are GMO foods 'substantially equivalent' to non-GMO foods, as the FDA maintains?' GMO FAQs. <https://gmo.geneticliteracyproject.org/FAQ/gmo-foods-substantially-equivalent-non-gmo-foods-fda-maintains/>

Goldy R (2019) 'New label denoting bioengineered ingredients will soon appear on food items in 2020'. Michigan State University Extension, 4 December 2019. <https://www.canr.msu.edu/news/new-food-label-denoting-bioengineered-ingredients>

Gouws LM, Groenewald J-H (2015) 'GM food labelling – is there a solution to the impasse?' *South African Food Science and Technology (FST) Magazine* July 2015. <https://www.academia.edu/37749097/GM_food_labelling_is_there_a_solution_to_the_impasse>

Library of Congress (2014) *Restrictions on Genetically Modified Organisms: Canada*. <https://www.loc.gov/law/help/restrictions-on-gmos/canada.php>

Lusk JL, McFadden BR, Wilson N (2018) Do consumers care how a genetically engineered food was created or who created it? *Food Policy* **78**, 81–90. doi:10.1016/j.foodpol.2018.02.007

Lynas M (2018) 'Brazilian debate highlights need for smart GMO labeling regime'. Alliance for Science, 30 April 2018. <https://allianceforscience.cornell.edu/blog/2018/04/brazilian-debate-highlights-need-smart-gmo-labeling-regime/>

McCluskey JJ, Wesseler J (Eds) (2018) Special issue on the economics and politics of GM food labelling. *Food Policy* **78**, 1–100.

National Academies of Sciences, Engineering and Medicine (2016) *Genetically Engineered Crops: Experiences and Prospects*. The National Academies Press, Washington, DC. <https://doi.org/10.17226/23395>

OECD (1993) *Safety Evaluation of Foods Derived by Modern Biotechnology: Concepts and Principles*. Organisation for Economic Co-operation and Development, Paris, France.

Schauzu M (2000) The concept of substantial equivalence in safety assessment of foods derived from genetically modified organisms. *AgBiotechNet* **2**.

Tosun J, Schaub S (2017) Mobilization in the European public sphere: the struggle over genetically modified organisms. *The Review of Policy Research* **34**, 310–330. doi:10.1111/ropr.12235

USDA (2018) *National Bioengineered Food Disclosure Standard.* <https://s3. amazonaws.com/public-inspection.federalregister.gov/2018-27283.pdf>

Yeh DA, Go'mez MI, Kaiser HM (2019) Signaling impacts of GMO labeling on fruit and vegetable demand. *PLoS One* **14**, e0223910. doi:10.1371/ journal.pone.0223910

Zilberman D, Kaplan S, Gordon B (2018) The political economy of labelling. *Food Policy* **78**, 6–13. doi:10.1016/j.foodpol.2018.02.008

9

The West versus Africa

The global West often refers to Africa as if it were a homogeneous set of countries, most of which, with the possible exception of South Africa, are considered to be 'basket cases'. That this is not true, certainly for agricultural biotechnology, has already been seen in Chapter 4, which presented the many sophisticated GM crops that are being developed by Africans for Africans. In this chapter we will consider the problems that have arisen in bringing these crops to the farmers who need them. Many of these problems have been fuelled by attitudes towards GM crops held by groups operating in the global West, particularly those in Europe.

As discussed previously, only South Africa, Sudan and, very recently, Eswatini and Nigeria are currently commercialising any GM crops. Why is this and what is preventing the uptake of GM crops by the rest of the continent? To understand the influence that Europe could be having in Africa, it is important to understand the role that Europe plays in both the economy and mindset of many African countries. A statement by the European Commission reads: 'In an ever-changing world, one thing is sure: Africa and Europe will remain each other's closest neighbours. Africa's 54 countries and the European Union's 28 Member States have a shared neighbourhood, history and future.' The article goes on to state that the Joint Africa–EU Strategy has a budget for the period 2018–20 of 400 million euros (European Commission 2019).

Much of Africa's agricultural produce is destined for Europe. Indeed, as Robert Paarlberg writes in his book *Starved for Science*, 'Africa's farm exports to Europe are six times as large as exports to the United States, so it is European consumer tastes and European regulatory systems that Africans most often must adjust to.' In addition,

Europe provides three times the funding for the United Nations Environment Program (UNEP) than does the USA. UNEP, together with the Global Environment Facility, provided assistance to African regulatory authorities. Therefore, Europe could influence organisations to adopt EU-style restrictions on GM crops and the EU has been waging a war on GMO foods for decades (Paarlberg 2008).

As discussed in previous chapters, the reasons for these attitudes are varied, but in an article published in the magazine *Reason*, entitled 'Europe's anti-GMO stance is killing Africans', Marion Tupy argues that, in reality, the EU is trying to protect its own farmers from competition with their possibly more productive American competitors. 'Thus, were the US food aid inadvertently to 'contaminate' Africa's crops, Africans would be in trouble' (Tupy 2017). As a result, not only does the EU's stand on GM crops affect Africa's exports, it also affects countries' ability to accept food aid from the USA.

This was presumably behind the stance taken by the Zambian government, during the crippling drought of 2002, when they refused food aid from the US Agency for International Development, via the World Food Program (WFP). The reason for this action was that the aid could possibly contain GM food, which, according to Zambia's president, Levy Mwanawasa, is 'poison'. Even after sending a delegation to South Africa, Europe and the USA to assess potential harm, Zambia's agricultural minister stated, 'In the face of scientific uncertainty the country should thus refrain from action that might adversely affect human and animal heath, as well as harm the environment' (Knight 2002).

Commenting on this situation in the South African *Sunday Times* newspaper of 10 November 2002, the journalist Mathatha Tsedu wrote that the Zambian president, Levy Mwanawasa, had said that he wanted to protect his people from poisoning. 'Very commendable indeed' writes Tsedu, 'save for the fact that the people he wants to save from death by poisoning are now dying from hunger.' He went on to say, 'The problem with many of this continent's leaders is that, once ensconced in office, they are so far removed from the reality of normal people's everyday lives it hurts' (Tsedu 2002).

During October of that year, the WFP was only able to supply food to approximately half of Zambians suffering from food shortages. Other African countries, also suffering from drought at that time, agreed to receive food aid as long as it was milled 'so that it cannot be planted and contaminate indigenous crops', presumably maize, which, at the time, could not grow because of the lack of water (Knight 2002).

The current status is that although GMOs are not barred from entering Europe – indeed, as discussed in Chapter 3, the EU is heavily dependent on GM soybeans for animal feed – their labelling legislation for any product containing at least 0.9% GM content is seen by African farmers as prohibitive. As a result, most African countries refuse to grow GM crops.

In an article published in 2017, Justus Wesseler, an agricultural economist from Wageningen University, and colleagues looked at the cost of delaying the introduction of three GM crops, namely disease-resistant cooking bananas (matoke) and insect-resistant cowpea and maize. 'If Kenya had adopted GE [genetically engineered] corn in 2006,' the study estimates, 'between 440 and 4,000 lives could theoretically have been saved. Similarly, Uganda had the possibility in 2007 to introduce the black sigatoka resistant banana, thereby potentially saving between 500 and 5,500 lives over the past decade' (Wesseler *et al.* 2017).

In addition, they expressed concern over the delay in approval of *Bt* cowpea by Nigerian officials, as they estimate that 'A one-year delay in approval [of the insect-resistant *Bt* cowpea], would especially harm Nigeria, as malnourishment is widespread there ... [and] cost Nigeria about 33 million USD to 46 million USD and between 100 and 3,000 lives.'

However, since that article appeared, and as will be discussed further below, Nigeria has approved the commercial release of *Bt* cowpea that is resistant to the pod borer, *Maruca vitrata*. Commenting on this announcement, Professor Ishayaku, the project's Principal Investigator, urged other African countries to stop raising problems about the use of such GM crops when such problems have already been addressed numerous times. He wrote, 'Regulators should stop the attitude of re-inventing the wheel: we have global and universal principles. What is applied in Rwanda can be applied in Kenya as long as it is healthy and environmentally friendly … [Regulators] should bear in mind that the financial burden through regulations reduces the profitability of the technology' (ISAAA 2019a). In other words, why don't African countries learn from each other and not begin afresh each time?

In opposition to the uptake of GM crops by African countries, an article entitled 'Twelve reasons for Africa to reject GM crops' was written by Zachary Makanya, who works for the PELUM (Participatory Ecological Land Use Management) Association. He wrote, 'Africa is in danger of becoming the dumping ground for the struggling GM industry and the laboratory for frustrated scientists. The proponents of GM technology sell a sweet message of GM crops bringing the second green revolution and the answer to African hunger, but a closer look makes it clear that GM crops have no place in African agriculture' (Makanya 2004).

On their webpage (https://www.pelum.net), it states that the PELUM Association is a network of 170 non-governmental organisations (NGOs) in 10 countries of East and Southern Africa. They work with small-scale farmers to improve their livelihoods through ecological land use and management. However, their members, or associate members, include Earth Greenery Activities Japan, which promotes organic farming practices in Tanzania and Japan, the German Appropriate Technology Exchange, MISEREOR, an overseas development agency of the Catholic Church in Germany, and NOVIB, a Dutch agency supporting grass roots development, and the Ford Foundation in the USA.

One of their arguments is that African farmers will lose their indigenous crop varieties and become reliant on multinational corporate takeovers of the local seed sector. In answer to this, writing on the situation in Uganda for the Alliance for Science's online newsletter, Isaac Ongu explains that these issues were prevalent ~60 years ago, long before GM crops became an issue. This was due to the introduction of hybrids by foreign companies, which led to farmers, impressed by improvements in yield and other useful traits, to start to buy new seeds every year (Ongu 2019). As discussed in Chapter 6, traits introduced by GE only add to the already useful traits present in pre-existing hybrids. For instance, in Uganda, the Ministry of Agriculture's National Crop Variety List, published by The African Seed Access Index (TASAI), shows that the first hybrid variety was introduced into that country in 1960 and by 2017 89 new maize varieties had been released to its farmers (TASAI 2017). In fact, most of the so-called 'local varieties' currently being grown are the products of scientific research that began before Uganda obtained its independence in 1962.

I well remember being on a field trip in the neighbouring country, Kenya, some years ago and listening to a woman farmer describing, through an interpreter, the added value of the pigeon peas she was growing that had been improved by the International Crops Research Institute for the Semi-Arid Tropics. Pointing derogatorily to a small patch of poorly growing plants nearby, she said that she would never plant local varieties again.

Of the 89 new maize varieties mentioned above, Ongu (2019) states that 38 (42.7%) were developed by foreign companies. The rest were based on research carried out in Uganda's public research system. Furthermore, the fact that multinational seed companies trade in Uganda is a result of the country's own Public Enterprise Reform and Divestiture policy of 1993 (Uganda Legal Information Institute 1993). Its objectives are to promote the development of an 'efficient, market-led private sector'. This enabling environment allowed direct foreign investments, such as seed companies, in the country.

In his article mentioned above, Ongu wonders whether 'GMO opponents who claim Uganda's new biosafety bill would open the country to multinationals like Monsanto are possibly ignorant of the fact that Ugandan farmers have been buying seeds from these [multinational] companies for over a decade' (Ongu 2019).

Another commentary on the role of European countries on the uptake of GM crops comes from Margaret Karembu, Director of the ISAAA regional office in Africa. She talks about her early years growing up in rural Kenya when it was a struggle to put food on the table for her family. She now realises that what her family was practising was subsistence farming, which European 'greens' call 'agro-ecology family farming', in which families hardly produce enough food to last until the next harvest. As a result, most farmers were locked into unsustainable food production that perpetuated poverty (Karembu 2017).

She further writes, that although farming practices are beginning to modernise, this is being undermined by other countries that appear determined to prevent Africa from joining the global agricultural revolution. One example is the adoption by the European Parliament in June 2016 of a report by the New Alliance for Food Security and Nutrition, which stated that any support to African agriculture should be confined to the aforementioned 'agro-ecology family farming level'. The report, passed by 577 to 24, attacked ongoing efforts by Africans to introduced advanced agricultural technologies, including GM crops.

The article quotes the 'Heubach Report' put forward by Maria Heubach, a German Green MEP, in which it is written: 'We have already made the mistake of intensive agriculture in Europe. We should not replicate it in Africa because this model destroys family farming and reduces biodiversity.' The Report goes on to state that the introduction of certified seeds in Africa increases smallholder dependence, makes indebtedness more probable and erodes seed diversity. The 'Group of Seven' (G7) is urged not to support GE crops in Africa. Karembu writes in rebuttal: 'I have kept in touch with my village roots and can authoritatively challenge this colonial mindset about Africa ... We are not an agricultural backwater as European politicians seem to believe but potentially the world's future food basket. More and more farmers appreciate the value of using certified seeds, which greatly out yield farm-saved seeds ... This has not in any way prevented those farmers who want to use farm-saved seed from doing so or undermined seed diversity' (Karembu 2017). In addition, Europe has some of the most stringent commodity import standards that have long since pushed small-scale suppliers from Africa out of the EU market in favour of large farms.

Karembu ends her article by citing the role that Western NGOs play in slowing down the progress of modern biotechnology in Africa. She refers to organisations such as Greenpeace, Friends of the Earth, GeneWatch UK, ActionAid and GM Freeze and their affiliates in

Africa, which claim that GM crops will undermine smallholder farmers and expose their populations to potential health risks. As she states, 'No health or science agency in the world has documented links between GM foods and any health hazard' (Karembu 2017).

In another article, entitled 'Suppressing growth: how GMO opposition hurts developing nations', the authors (Giddings *et al.* 2016) point out that farmers in developing countries can often not afford many of the innovations that boost agricultural productivity. These include modern tractors, fertilisers, herbicides, pesticides and crop sprayers etc., but improved seeds they can more easily afford. This is why farmers in developing countries are planting more biotech-improved seeds than those in industrial countries (ISAAA 2018).

The involvement of Sweden's anti-GMO community in African agriculture was highlighted by Per-Ola Olsson, a Swedish journalist and educator, in an article entitled 'Examining Sweden's ties to anti-GMO conference in Africa, through taxpayer-funded Swedish Society for Nature Conservation'. He was referring to the 1st International Conference on Agroecology, which was held in Nairobi, Kenya in June 2019, where one of the main contributors was the Swedish Society for Nature Conservation (SNF). SNF receives approximately US$9 million a year from the Swedish government and from the Swedish International Development Cooperation Agency. Some of the money donated by SNF was used to sponsor speakers such as Gilles-Éric Séralini, whose paper purporting to show that GM maize causes cancer was withdrawn (see Chapter 6), Judy Carman, whose study showing a link between GM maize and inflammation of the stomach of pigs, was contradicted by field data showing no difference in the health of animals fed GM or non-GM feed over 18 years since 1996, and other anti-GMO speakers (Olsson 2019).

Now let us consider some of the countries in Africa that have been targeted by anti-GMO campaigners, operating both from within and outside their countries, over the past few years.

Kenya

The publication in 2012 of the paper by French scientist, Gilles-Éric Séralini, entitled 'Long term toxicity of a Roundup herbicide and a

Roundup-tolerant genetically modified maize' (Séralini *et al.* 2012), had major consequences for the use of GM crops in Kenya. On 21 November 2012, the Kenyan Ministry of Public Health ordered the removal of all GM foods on the market and banned the importation of GM products. The Minister for Public Health, Beth Mugo, herself a cancer survivor, presented her concerns to President Kibaki following the Séralini publication. The President accepted her recommendation and decreed the ban without consulting the National Biosafety Authority (NBA). The NBA, under the Ministry of Higher Education, Science and Technology is mandated to 'exercise general supervision and control over the transfer, handling and use of genetically modified organisms with a view to ensure safety of human health and provision of adequate protection of the environment' (Snipes and Kamau 2012).

Over the years since 2012, the President has often given indications that the ban could be rescinded and the latest information came in October 2019. In an article published by AgriBusiness Global (Njiraini 2019), Dorington Ogoyi, CEO of Kenya's NBA stated, 'The Séralini study that influenced the ban has widely been discredited, and, as a country, we feel lifting the ban is prudent in order to fully benefit from the GM technology.'

The decision to overturn the ban was influenced by the three studies funded by the EU with the aim of ascertaining the safety of GMOs to humans and the environment. These studies, GRACE, GMO90+ and G-TwYST, showed that glyphosate-resistant GM maize (NK603 in particular) does not pose any health risks. In addition, the latter (G-TwYST: GM Plant Two Year Safety Testing) concluded that there was no evidence of carcinogenicity in rats or maize using either NK603 or NK603 treated with glyphosate. 'It was concluded that there were no adverse effects related to the administration of the GM maize NK603 cultivated with or without Roundup,' the report states (G-TwYST 2018).

Lifting the ban will mean that Kenya can import GM food and products, especially cheap maize, from countries such as Brazil and South Africa, to cover shortfalls in production caused by drought and other problems. On average, Kenya imports ~5 million bags of maize (Njiraini 2019).

In addition, waiting in the wings for approval, are *Bt* cotton and *Bt* maize. The Kenya Agriculture and Livestock Organization (Kalro) has been conducting national performance trials of *Bt* cotton at seven sites across the country. According to Dr Charles Waturu, Kalro's Principal Researcher for GM cotton, these trials yielded more than four times what farmers had been harvesting using local varieties. Kalro is also carrying out research on *Bt* maize and has been waiting for the NBA to give them permission to undertake performance trials (Njiraini 2019). Finally, in December 2019 the Kenyan Cabinet, chaired by President Uhuru Kenyatta, approved the commercial planting of *Bt* cotton. This came after the 5-year field trials showed positive results and seeds will be available by March 2020. By doing this, Kenya aims to become a leading global player in textile and apparel production (ISAAA 2019b).

This section on Kenya cannot end without mention of a remarkable organisation based in Nairobi. It is the African Agricultural Technology Foundation (AATF; www.AATF-Africa.org), which was mentioned briefly in Chapter 4. It was founded in 2003 and I was the first Chair of the Board. Its aim is to provide farmers in subSaharan Africa with access to technological solutions to improve productivity and improve food security. It often obtains technology from multinational companies for farmers and seed suppliers in the region to use royalty-free. Among its success stories are, as discussed in Chapter 4, *Bt* cowpea and drought-tolerant/insect-resistant maize. It thus seems ironic that the very country bringing new GM crops to subSaharan Africa has, until very recently, not been allowing its own farmers access to this technology.

Uganda

Uganda has been struggling for many years to have a bill passed in parliament, and signed into law by the President, that will allow the country to commercialise GM crops. In 2013 the National Biotechnology and Biosafety Bill was introduced in parliament, but it took 4 years, until October 2017, to be approved as the *Biosafety Act*, only to be rejected by the President. Many activists, some close to his office, contested it because they characterised GM crops as harmful, enslaving and risky (Wamboga-Mugirya 2019).

One of the reasons given by President Moseveni was that some clauses did not favour livestock farmers. He also questioned the patent rights of indigenous farmers, claiming also that scientists were integrating GMOs with indigenous crops and animals. As a result, the law was returned to parliament, which made adjustments and passed it again as the *Genetic Engineering Regulatory Act* (GERA) in November 2018. While waiting for a decision by the President, there were concerns over the new liability provisions that carried potentially harsh penalties for researchers. Although the previous version of the Act had included a fault-based liability provision, the current Clause 35 states that in the event of any harm to the environment or human health, any person or company holding a patent for that product is presumed to be guilty. This strict liability clause could potentially negatively affect local scientists working for state-owned research organisations or universities (Wamboga-Mugirya 2019).

While waiting for the passage of the GERA, Professor Joseph Obua, chairman of the Governing Council of the country's National Agricultural Research Organization, urged legislators to reject the strict liability clause and instead adopt the previous fault-based one. He expressed concern that Clause 35 would criminalise local scientists, saying that there are existing legislations with liability regimens for managing other plant breeding methods and these do not provide strict liability for errors and commissions (Wamboga-Mugirya 2019).

In the meantime, the President refused to sign the Act, this time stating that it did not favour the inventor and the farming community. He wanted the legislators to 'include clauses that will guarantee equal sharing of GMO products between the breeders and vendors, provide assurances that GMO and non-GMO seeds will not be mixed, and detail forms of genetic modification, including use of poisonous substances or bacteria, when developing GMO products' (Afedraru 2019).

It is possible for parliament to enact the bill without the President's signature, because of the length of time that it has taken for this to happen. Uganda's constitution states that under these circumstances the Speaker can allow a bill to become law if it is supported by at least two-thirds of all MPs.

However, there is still concern about the strict liability clause and some scientists are arguing that this should first be amended. They say that they will continue working to educate policy makers and all Ugandans on the value of modern biotechnology (Afedraru 2019).

Tanzania

In 2016, the Tanzanian Agricultural Research Institute (TARI) began confined field trials of GM maize in Makutopora in the Dodoma Region, having had to wait 5 years for permission after they had applied for it to the country's regulatory authority. TARI, a semi-autonomous body of the government under the Ministry of Agriculture, is responsible for conducting all agricultural research activities in the country. They carried out the first trials to have been approved in Tanzania, in order to compare conventionally bred water-efficient maize with GM maize developed under the Water Efficient Maize for Africa (WEMA) project. This project is managed by the AATF and uses a gene encoding a cold-shock protein, *cspB*, from the bacterium *Bacillus subtilis*, which had been shown to confer dehydration tolerance to maize (Castiglioni *et al.* 2008). The gene was donated royalty-free by Monsanto and introduced into local varieties of maize by CYMMIT, the international maize and wheat breeding organisation, which, although based in Mexico, has centres in Africa. It was these plants that were undergoing approved field trials in Makutopora.

Imagine, therefore, the shock of members of TARI when, in November 2018, the permanent secretary of the Ministry of Agriculture, Mathew Mtigumwe, announced not only a ban on the ongoing trials, but also directed TARI to immediately destroy all evidence of their research (Mirondo 2018). Mirondo quotes Mr Mtigumwe as stating: 'TARI has recently invited various groups to the Makutopora centre to witness its research findings when the government is yet to approve use of GM products in the country.'

It would appear that actions taken by TARI, which included making public statements about the trials, annoyed parliament. The then newly appointed Minister of Agriculture, Japheth Hasunga, told *The Citizen* newspaper that TARI had contravened government approval processes. 'They were supposed to give my ministry the findings ... [so that we in turn could have] consulted with other ministries to satisfy ourselves that the said GM seeds were safe and did not carry any risks to humans', said Mr Hasunga (Mirondo 2018). Researchers at TARI expressed surprise that a technical misstep was the reason for the decision that would deny farmers access to seeds with traits that could solve local production challenges, including drought and insect attack.

The action taken by the Ministry of Agriculture was endorsed by the French anti-GMO website, Inf'GMO, which reported, as quoted in the Genetic Literacy Project by Brazeau, after English translation (Brazeau 2018): 'This decision was taken following the publication by TARI of the results of these tests without having obtained the necessary authorization. TARI has indeed communicated widely on the success of these tests and organized various lobbying actions, including inviting the Parliamentary Committee for Food and Agriculture to visit its facilities. But the tests are not supposed to be open to the public. The ministry holds that by organizing these lobbying actions, TARI went beyond its mission. Minister Hasunga also denounced the description by some pro-GMO activist researchers of smallholder peasants as 'poor and hungry' for pro-GMO propaganda purposes.'

The French article goes on to say, in the English translation, that 'The WEMA program is also seeking to weaken the current framework, in particular by asking for a change in liability for damage caused by these crops. Currently, the standard applied is that of strict liability. The WEMA program would like to replace that with fault-based liability. Strict liability means that anyone who introduces GMOs into the environment is directly liable for any damage or harm while fault-based provisions mean that the fault or negligence of anyone who introduces GMOs must first be proven.'

The reason for the ban of the trials and the destruction of research findings is, as stated by Mr Husanga, indeed surprising. It could make sense to sanction, in some way, the TARI scientists involved in the trials and their public statements about them, but it makes no sense to go to the lengths of destroying their work over an apparent neglect of protocol.

The current situation in Tanzania appears to be confusing. At the same time as the maize trials were being carried out by TARI, trials on GM cassava were also underway at the Mikocheni Agricultural Research Institute (MARI) in Dar es Salaam, but they have apparently not been stopped or destroyed. In June 2019, Dr Freddy Tairo, the country coordinator of Biotechnology Research at MARI, stated that research on cassava was continuing at their laboratory and that their findings would be submitted to the government for further action (Abdu 2019).

Nigeria

In 2019 Nigeria approved the release of two GM cotton varieties containing the *Bt* gene to protect the crop from bollworms. Cotton was once a driver of the country's textile industry, employing ~350 000 people. In later years the sector became moribund due to low yields caused both by drought and by pests. The government hoped that the new GM cotton varieties would increase the yield and thereby cut the cost of imports which amounted to as much as US$319 million a year (Olurounbi 2019).

Soon afterwards, the Nigeria Biosafety Management Agency (MBMA) approved the commercial release of *Bt* cowpea, the first country in Africa to do so. Pod borer-resistant cowpea, mentioned in Chapter 4, is resistant to the boring insect *Maruca vitrata*, which can cause 70–90% yield losses for farmers. Confined field trials were carried out by the Institute for Agricultural Research and showed that *Bt* cowpea could reduce the use of insecticides from eight sprays per season to about two and increase yield by up to 20% (IITA 2019). Then, on 12 December 2019, as described in Chapter 4, the MBMA approved the registration and commercial release of variety called 'SAMPEA 20-T'. This one, as well as being resistant to the pod borer, is also high yielding, early maturing and resistant to two parasitic weeds, *Striga* and *Alectra* (ISAAA 2019c).

However, anti-GMO activists in Nigeria are making their opposition to these releases felt, despite the fact that *Bt* cowpea was developed by the country's own scientists in collaboration with the

AATF. Cowpea is a major source of protein to both rural and urban poor in West Africa and Nigeria is the largest producer. Increasing productivity of this important crop will therefore bring considerable relief to the country's poorest inhabitants.

Despite this, many anti-GMO activists have spoken out against the approvals of GM crops in general. For instance, Health of Mother Earth Foundation (HOMEf) organised a dialogue with students and staff from the Faculty of Agriculture at the University of Benin on 30 April 2019. One of the speakers was Mariann Bassey-Orovwuje, who is the coordinator of the food sovereignty programme for Friends of the Earth Nigeria/Africa. She stated that GMOs will foster corporate control of food systems, destroy Nigeria's biological diversity, lead to an irreversible contamination of indigenous seed varieties and loss of local knowledge (HOMEf 2019).

It should be noted that Friends of the Earth, which started in the USA in 1969, became an international organisation in 1971 when it was joined by Sweden, the UK and France. Their homepage states: 'Friends of the Earth works to transform public policy to establish appropriate safety assessment and oversight of GMO crops and animals, and we lead campaigns to keep poorly regulated GMOs out of our food system.' Their representative at the meeting at the University of Benin was clearly of the view that the Nigerian regulatory authorities are not capable of regulating GM crops.

It will be interesting to see how the new variety of GM cowpea is received by Nigerian farmers and how the anti-GMO organisations respond.

Eswatini

This landlocked country, surrounded by southern Africa, was previously called Swaziland. More than 10% of its 1.4 million inhabitants are engaged in farming, many in cotton. In an interview with the Alliance for Science (Isaac 2019), Dr Daniel Khumalo, the CEO of the Swaziland Cotton Board, explained that this crop has a long history in the Kingdom, serving as a source of livelihood for over 50 000 Emaswati, as the inhabitants of Eswatini are known. 'The country's richest people used to be cotton farmers who owned cars, tractors, good houses … with the money from cotton'. However, due to the crippling drought in recent years the price of cotton has dropped, resulting in the closure of the ginnery. Already in 2012, the country had passed the *Biosafety Act* and in 2018 released *Bt* cotton. Commenting on this event, Dr Khumalo went on to say, 'I will advise countries to review their legislations in order to benefit from products of modern biotechnology. To those with accommodating legislations, I want to urge them to put their political ego aside and think of the poor African farmers who are striving in the field with old age technology when the world has invented new technologies that would benefit them.'

Furthermore, he noted, 'Africa will never be food secure until the continent adopts technologies that will improve production. GM technology is one solution Africa needs to consider.' In just a single season GM cotton has led to a doubling of the ginnery's consumption.

The opposition was not slow to follow up on these new initiatives and in an open letter to the Cotton Board, PELUM (mentioned before as one of the anti-GMO groups active in Africa), asked several questions (PELUM 2018). However, in this case, the questions were justified and it will be interesting to hear the answers once the cotton crop has been analysed. For instance, 'What percentage of the field was saved for refuge?'; 'Who is responsible for farmer training and certification?' 'How many farmers have been trained and certified?'; 'How many field inspections have been conducted and what were the conclusions of those inspections'; 'Has there been a decrease in the number of pesticides sprayings of the fields compared the non-*Bt* cotton fields?'; and finally, 'How will smallholder farmers be able to afford the irrigation that is necessary? How will they be able to afford to buy the BT cotton seeds every year or pay for the pesticides or fertilisers it requires?'

However, even though these questions have not yet been answered, PELUM already concludes that 'Growing BT cotton isn't economically feasible for Eswatini smallholder farmers.' This is just the attitude that Dr Khumalo was referring to when he stated that Africa will never be food secure until the continent adopts technologies that will improve production. GM technology is one solution Africa needs to consider

(Isaac 2019). In other words, don't condemn any technology without allowing farmers to first put it to the test. Farmers are savvy people and will not continue to use a technology that is not in their best interests.

In conclusion, how have Western attitudes affected the uptake of GM crop production in Africa? First, as Europe provides major funding to Africa and much of Africa's agricultural produce is destined for Europe, the attitude of Europeans towards GM crops will undoubtedly influence Africans, especially their leaders and decision makers, and it is well known that the EU has been waging a war on GMO foods for decades. The fact that the EU imports GM soybeans to feed their animals, but does not allow their farmers to grow this crop, does not appear to be ironic to Europeans.

Second, some Western-based anti-GMO organisations use the argument that Africa will become a dumping ground for unwanted GM crops. This has been an argument since the beginning of the development of GMOs in the 1980s. On the contrary, this argument is highly insulting towards African governments, implying that they are unable to manage their regulatory authorities.

Third, many in the West would like to keep farms in Africa as 'family farms', presumably only feeding their own members. The mere thought of Africans becoming commercial farmers appears to horrify them. Linked to this is the attitude that African farmers should rely on farm-saved seeds and not plant improved varieties or certified seeds. In fact, as stated above, Europe has some of the most stringent commodity import standards, which are effectively keeping small-scale suppliers from Africa out of the EU market in favour of large commercial farms.

Fourth, on the question of financial security, farmers in Africa can often not afford many of the modern agricultural innovations, such as fertilisers, herbicides and insecticides, to say nothing of modern tractors, crop sprayers etc. However, one thing they can more readily afford is improved seeds.

Finally, there are the totally unfounded statements by many Westerners that GM crops are harmful to human health and can even cause cancer. I recently met two African PhD students studying in Tasmania, one in medical science. I gave them a copy of my most recent book on GM crops entitled *Food for Africa*. To my amazement they recoiled from me in horror. 'But GM crops give you cancer!' they both exclaimed. Séralini has a lot to answer for.

The next chapter will consider how other countries in the developing world have been affected by attitudes held in the West.

References

Abdu F (2019) 'GMO research not prohibited in Tanzania – expert'. *Daily News*, 29 June 2019. <https://www.dailynews.co.tz/news/2019-06-295d1719a911818.aspx>

Afedraru L (2019) 'Plans to introduce GMO crops in disarray, legislators angry after Uganda's president rejects GMO cultivation law for second time'. Genetic Literacy Project, 9 September 2019. <https://geneticliteracyproject.org/2019/09/09/plans-to-introduce-gmo-crops-in-disarray-legislators-angry-after-ugandas-president-rejects-gmo-cultivation-law-for-second-time/>

Brazeau M (2018) 'Why did Tanzania just pull the plug on its GMO crop trials?' Genetic Literacy Project, 28 November 2018. <https://geneticliteracyproject.org/2018/11/28/why-did-tanzania-just-pull-the-plug-on-the-its-gmo-crop-trials/>

Castiglioni P, Warner D, Bensen RJ, Anstrom DC, Harrison J, Stoeker M, Abad M, Kumar G, Salvador S, D'Ordine R, Navarro S, Back S, Fernandes M, Targolli J, Dasgupta S, Bonin C, Luethy MH, Heard JE (2008) Bacterial RNA chaperones confer abiotic stress tolerance in plants and improved grain yield in maize under water-limited conditions. *Plant Physiology* **147**, 446–455. doi:10.1104/pp.108.118828

European Commission (2019) *Africa–EU Continental Cooperation*. <https://ec.europa.eu/europeaid/regions/africa/africa-eu-continental-cooperation_en>

G-TwYST (2018) *Conclusions and recommendations*. <https://www.g-twyst.eu/files/Conclusions-Recommendations/G-TwYSTConclusionsandrecommendations-final.pdf>

Giddings LV, Atkinson RD, Wu JJ (2016) 'Suppressing growth: how GMO opposition hurts developing nations'. Information Technology and Innovation Foundation, 8 February 2016. <https://itif.org/publications/2016/02/08/suppressing-growth-how-gmo-opposition-hurts-developing-nations>

HOMEf (2019) 'Students of agriculture reject GMOs; call for protection and promotion of indigenous farming systems'. HOMEf, 3 May 2019. <https://homef.org/2019/05/03/students-of-agriculture-reject-gmos-call-for-protection-and-promotion-of-indigenous-farming-systems/>

IITA (2019) 'Major breakthrough for farmers and scientists as Nigerian biotech body approves commercial release of genetically modified cowpea'. International Institute of Tropical Agriculture, 9 February 2019. <https://www.iita.org/news-item/major-breakthrough-for-farmers-and-scientists-as-nigerian-biotech-body-approves-commercial-release-of-genetically-modified-cowpea>

ISAAA (2018) 'Global status of commercialized biotech/GM crops 2018: biotech crops continue to help meet the challenges of increased population and climate change'. ISAAA Brief No. 54. ISAAA, Ithaca, NY.

ISAAA (2019a) 'Prof. Ishayaku: Nigeria's approval of Bt cowpea signifies Africa's capacity to adopt biotech crops'. ISAAA, 13 February 2019. <http://africenter.isaaa.org/prof-ishayaku-nigerias-approval-bt-cowpea-signifies-africas-capacity-adopt-biotech-crops/>

ISAAA (2019b) 'Bt cotton approved for planting in Kenya'. Crop Biotech Update, 19 December 2019, <http://www.isaaa.org/kc/cropbiotechupdate/article/default.asp?ID=17902>

ISAAA (2019c) Nigeria commercializes pod borer resistant cowpea, its first GM food crop. Crop Biotech Update, 18 December 2019. <http://www.isaaa.org/kc/cropbiotechupdate/article/default.asp?ID=17899>

Isaac N (2019) 'Swaziland (eSwatini) finds success with GMO cotton'. Alliance for Science, 6 June 2019. <https://allianceforscience.cornell.edu/blog/2019/06/swaziland-eswatini-finds-success-gmo-cotton/>

Karembu M (2017) 'How European-based NGOs block crop biotechnology adoption in Africa'. Reposted from Genetic Literacy Project, 24 February 2017. <http://africenter.isaaa.org/european-based-ngos-block-crop-biotechnology-adoption-africa/>

Knight W (2002) 'Zambia bans food aid'. *New Scientist*, 30 October 2002. <https://www.newscientist.com/article/dn2990-zambia-bans-gm-food-aid/#ixzz5zxh7P756>

Makanya Z (2004) 'Twelve reasons for Africa to reject GM crops'. *Seedling* July 2004, 18–22. <https://www.grain.org/article/entries/427-twelve-reasons-for-africa-to-reject-gm-crops>

Mirondo R (2018) 'Shock as government bans GMO trials'. *The Citizen*, 1 November 2018. <https://www.thecitizen.co.tz/news/-Shock-as-government-bans-GMO-trials/1840340-4865040-12al5ps/index.html>

Njiraini J (2019) 'Kenya set to rescind GMO ban'. *AgriBusiness Global*, 1 October 2019. <https://www.agribusinessglobal.com/markets/africa-middle-east/kenya-set-to-rescind-gmo-ban/>

Olsson P-O (2019) 'Examining Sweden's ties to anti-GMO conference in Africa, through taxpayer funded Swedish Society for Nature Conservation'.

Genetic Literacy Project, 25 September 2019. <https:// geneticliteracyproject.org/2019/09/25/examining-swedens-ties-to-anti-gmo-conference-in-africa-through-taxpayer-funded-swedish-society-for-nature-conservation/>

Olurounbi R (2019) *Nigeria approves two GMO cotton varieties to increase output*. <https://www.bloomberg.com/news/articles/2019-05-08/nigeria-approves-two-gmo-cotton-varieties-in-bid-to-boost-output>

Ongu I (2019) 'Growth of Uganda's seed sector exposes major anti-GMO claims'. Alliance for Science, 5 February 2019. <https://allianceforscience.cornell.edu/blog/2019/02/growth-ugandas-seed-sector-exposes-major-anti-gmo-claims/>

Paarlberg R (2008) *Starved for Science: How Biotechnology is Being Kept Out of Africa*. Harvard University Press, Cambridge, MA.

PELUM (2018) *Annual Report*. <https://www.pelum.net/wp-content/uploads/2019/08/Pelum-2018..-Annual-Report__210619...pdf>

Séralini G-E, Clair E, Mesnage R, Gress S, Defarge N, Malatesta M, Hennequi D, de Vendômois JS (2012) RETRACTED: Long term toxicity of a Roundup herbicide and a Roundup-tolerant genetically modified maize. *Food and Chemical Toxicology* **50**, 4221–4231. doi:10.1016/j.fct.2012.08.005

Snipes K, Kamau C (2012) 'Kenya bans genetically modified food imports'. *GAIN Report*, 27 November 2012. <http://agriexchange.apeda.gov.in/MarketReport/Reports/Kenya_Bans_Genetically_Modified_Food_Imports_Nairobi_Kenya_11-27-2012.pdf>

TASAI (2017) National crop variety list for Uganda. <https://tasai.org/wp-content/themes/tasai2016/info_portal/Uganda/National%20Crop%20Variety%20List%20for%20Uganda%20(2015).pdf>

Tsedu M (2002) 'Zambian tragedy: in the land of the starving, the one-eyed king is blind'. *Sunday Times* (Johannesburg), 10 November 2002.

Tupy M (2017) 'Europe's anti-GMO stance is killing Africans'. *reason*, 9 May 2017. <https://reason.com/2017/09/05/europes-anti-gmo-stance-is-killing-afric/>

Uganda Legal Information Institute (1993) Public Enterprises Reform and Divestiture Act 1993. <https://ulii.org/ug/legislation/consolidated-act/98>

Wamboga-Mugirya (2019) 'Question of liability: why researchers are worried about Uganda's new biotech act'. Genetic Literacy Project, 31 January 2019. <https://geneticliteracyproject.org/2019/01/31/question-of-liability-why-researchers-are-worried-about-ugandas-new-biotech-act/>

Wesseler J, Smart RD, Thomson J, Zilberman D (2017) Foregone benefits of important food crop improvements in sub-Saharan Africa. *PLoS One* **12**, e0181353. doi:10.1371/journal.pone.0181353

10

The West versus the Rest

Having seen the effects that attitudes and regulations in the West are having on countries in Africa, it is time to turn our attention to other so-called 'developing countries'. The reason why I use quotation marks is that one hopes that all countries are developing and unfolding their potential to varying extents. So why label some as developing and some not? This conundrum, which is based on a country's gross domestic product, will be covered in more detail in Chapter 11.

Have Western attitudes had similar effects as in Africa? We will look at cotton in India and Pakistan, eggplants (also known as brinjals, aubergines or talong) in India, the Philippines and Bangladesh, Golden Rice in Asia in general and maize in Mexico. In addition, although it is not strictly speaking a developing country, but because it has some unique features, we will consider virus-resistant papayas in Hawaii. Finally, the issue of gene editing using the relatively new technology known as CRISPR will be discussed.

Cotton in India and Pakistan

In Chapter 6 I addressed whether it is reasonable to link farmer suicides in India with GM crops, but now a new problem has arisen that is being used by anti-GMO activists, both in the West and in India in particular, to vilify *Bt* cotton. The problem is the pink bollworm pest, which appears to be developing resistance to the insecticidal proteins produced by *Bt* cotton. For instance, the headline in a 2018 Bloomberg article (Kulkarni 2018) reads: 'As a genetic revolution collapses, Vidarbha's cotton farmers dread coming season.' It reports that in 2017 the Maharashtra region 'witnessed the worst crisis in the history of *Bt* cotton since the seed technology was approved in India in 2002.' Not only was the cotton harvest decreased due to damage by the pink bollworm pest, but the article alludes to numerous deaths due to the resurgence of pesticide spraying. (Of peripheral interest is the fact that this is one of the few articles published by the anti-GMO fraternity that acknowledges deaths caused by insecticide spraying.)

In an article that attempts to unravel the facts behind this dramatic headline, Lynas (2018a) notes that the Kulkarni article admits that the apparent failure of *Bt* cotton is 'unique to India' and not found in other *Bt*-growing countries. So, what is going wrong in India? One problem is that studies have shown that many farmers in that country who thought they were growing *Bt* cotton were, in fact, planting fake GM seeds.

For instance, an article published by *The Tribune* on 8 March 2018, under the heading 'Dealer booked for selling fake *Bt* cotton seed', stated that a Gujarat-based company was accused of allegedly selling spurious seeds to six farmers who subsequently suffered huge losses due to decreased yields (The Tribune 2018). Again, another article published by *The Hindu* on 29 June 2019, under the headline: 'Farmers at their wits' end over fake seeds', indicated that this problem has been going on since 2000 (Guntur 2019). Therefore, fake GM seeds could be adding to the problem of pink bollworm infestations. This problem certainly requires further investigation. It is all too easy for anti-GMO critics to point to a failure in the technology when all the time the crop is not *Bt* at all. Once a thought has been 'seeded' in the public mind it is very hard to dislodge it.

Another problem as identified by Ron Herring, a Cornell University expert on GM cotton in India and cited by Lynas (2018a), is that Indian *Bt* cotton farmers might be a victim of their own success, because many of them are abandoning the recommended rotation of a second crop, which can be less profitable than the cash-crop cotton. 'In some areas, cotton has become essentially a perennial crop, picked continuously', Herring said in the article. However, 'there are implications of this practice; the longer you leave cotton in the field, the more likely the pink bollworms will show up.' Herring considers that restoring rotations in order to break the breeding cycle of the pest might well help. This is a perennial problem among farmers worldwide, whether the seeds being planted are GM or not – success of one crop can result in monocultures.

Indeed, in support of Herring's contention, Dr Paranjape, the associate director of the USAID-funded *Bt* brinjal project in Bangladesh, and an expert on *Bt* cotton in India, agreed that the problem might be the continuous cropping of cotton in certain regions. He contrasted the situation in Maharashtra described above with the irrigated cotton in the north of the country. There the cotton crop tends to be rotated with winter rice, which 'breaks the life-cycle of the pest' (Lynas 2018a). In addition, as Herring pointed out, 'Vidarbha [the region of Maharashtra where the problem is greatest] is severely drought-prone and cotton often fails [there] regardless of pest resistance.' Once again, climate change ominously rears its head, but the failure of a GM crop can be used by anti-GMO activists to bolster their arguments regardless of the cause of that failure.

Returning to the subject of myths as discussed in Chapter 6, Lynas (2018a) concludes his article: 'These expert testimonies confirm that, as so often is the case in science – and indeed the real world – the true story tends to be more complicated than the myth. Unfortunately, these simplistic myths tend to proliferate on the internet, especially when they fit a certain ideological narrative.'

Despite all the voices, both from the West and from Indian critics such Vandana Shiva, that decry the growing of *Bt* cotton in India, the evidence at this stage shows that this crop is likely to stay. As long as farmers choose it over other varieties it will continue to be grown in India.

Interestingly, there has been very little criticism of *Bt* cotton in neighbouring Pakistan. Perhaps this is because it has been shown that yields on farms growing this crop are ~20% higher than those growing the conventional crop (Kouser and Qaim 2014a). This could also be linked with the finding that *Bt* adoption increases environmental efficiency by 37% (using the Environmental Impact Assessment mentioned in Chapter 5; Kouser and Qaim 2014b). A third factor could be the effect that the decrease in insecticide spraying has on human health. When studying this aspect, the authors (Kouser and Qaim 2019) did not calculate their results based on farmers' belief that they were planting *Bt* cotton, as a significant amount of so-called '*Bt* cotton seed' is, as discussed above, fake. Instead they measured adoption by the level of *Bt* gene expression on farms. Using a cost-of-illness approach, they determined that 'true-*Bt* seed adoption decreases farmers' health costs by 33%.' They extrapolated this estimate to the entire *Bt* cotton area in the country and found that this resulted in an annual health cost savings of approximately US$7 million.

Bt eggplants in the Philippines, Bangladesh and India

In an article entitled 'State science, risk and agricultural biotechnology: *Bt* cotton to *Bt* brinjal in India' (Herring 2015), the author analyses the conundrum that two crops carrying the same transgene and facing the same authorisation procedures produced entirely different outcomes. State science will be discussed in Chapter 11, but for this discussion it can be viewed as the way a particular state interprets the scientific data presented to it. Thus, in India, the state science that approved *Bt* cotton was attacked as biased and dangerously inadequate by opponents, but the technology spread to virtually universal adoption by farmers. *Bt* eggplants were approved by the same Genetic Engineering Approval Committee (GEAC) that approved *Bt* cotton, but the eggplant decision was overruled, the GEAC was restructured and a moratorium was imposed on the crop.

As explained by Herring in his article (Herring 2015), the conflicts surrounding *Bt* eggplants were taken up by international organisations such as Greenpeace, quoting work by Séralini (whom we met in Chapter 6) that purported to show the danger that *Bt* eggplants posed to the environment and human health (Greenpeace India 2009). This expanded the arena in which the subject was debated and resulted in the restructuring of the GEAC, as mentioned above. In this way the different pathways taken in India regarding the two *Bt* crops reflected not only global patterns, but also the way in which the crops are cultivated as well as the interests of the state. In the case of *Bt* cotton, state and farmer interests, largely financial, dominated over precautionary attitudes, whereas for *Bt* eggplants state precaution prevailed. In other words, an edible GM crop received different attention from non-edible ones.

Therefore, let us look at the development of *Bt* eggplants. Eggplants (*Solanum melongena* L.) are among the most important, inexpensive and popular vegetables grown and consumed in Asia. In the Philippines they account for more than 30% of the total volume of vegetables produced in the country (Hautea *et al.* 2016). However, the vegetable is susceptible to widespread infestation by the eggplant fruit-and-shoot borer (EFSB; *Leucinodes orbonalis* Guenée). The damage is due to the holes and feeding tunnels produced by the adult insects and the excrement left by the larvae. As biological control and manual removal of EFSB-damaged fruits and shoots has proven ineffective, almost all farmers use chemical insecticides to control the borers. Eggplant farmers in the Philippines apply these chemicals 20–72 times during the 5–6-month-long cultivation season. Residues of these insecticides can be detected on harvested fruits. Farmers and farm workers complain of skin irritation, redness of eyes, muscle pains and headaches linked to the chemicals (Del Prado-Lu 2015).

Conventional breeding has been unsuccessful in producing commercial varieties of eggplants with resistance to EFSB. Therefore, scientists at Maharashtra Hybrid Seeds (Mahyco) turned to a GM solution, which they tested in India, Bangladesh and the Philippines. Hautea *et al.* (2016) published their results after analysing the field performance of Mahyco variety EE-1 in the Philippines. They found excellent control of EFSB shoot damage at levels of 98–100%. Fruit damage was diminished by 98–99% and larval infestation by 96–99%. In 2015, after this work had been carried out, but before its publication, the Supreme Court of the Philippines placed a permanent injunction on the conducting of field trials of *Bt* eggplants. This injunction, however, did not last very long, because in 2016 the same court unanimously set it aside. They gave as their reason that the case should have been dismissed because the field tests had been completed and there was therefore nothing to stop (Conrow 2016). In doing this, the Supreme Court dismissed on the grounds of mootness the petition filed by Greenpeace (Philippines). Thus, although this was not a total victory for the *Bt* eggplant in the Philippines, at least the research could carry on even though there is, as yet, no sign of any commercial approval.

In India, as mentioned above, the biosafety regulatory agency gave approval to the Mahyco variety EE-1 in 2009, but the Ministry of the Environment and Forests placed a moratorium on its cultivation in

2010 (Hautea *et al.* 2016) which is still in place. However, the anti-GMO activists are still monitoring the situation and, in an article entitled: 'Damning allegations: India's fields still have *Bt* brinjal', the Coalition for a GM-Free India alleges to have found the banned GM crop in the fields of a farmer in Haryana's Fatehabad region of the country (Sushma 2019). Representatives of the Coalition demanded immediate action from the central and state governments. This despite the fact that the farmer in question was happy with his crop.

There is a twist in this story, however. In June 2019 farmers belonging to a farmers' organisation called Shetkari Sanghatana, gathered in a village in Maharashtra where they symbolically planted *Bt* eggplants in defiance of the government moratorium. They called their action 'satyagraha', a term that was coined by the Indian independence hero Mahatma Gandhi for non-violent civil disobedience campaigns against the unjust laws imposed by the colonial regimen. The action by the farmers came about after the Indian authorities destroyed *Bt* eggplant crops that had been planted in Haryana in defiance of the moratorium. The protesting farmers demanded the government compensate the farmers whose crops had been destroyed and stated that 'in solidarity, the participants are raising money to aid all those who may be similarly affected' (Alliance for Science 2019). The matter is still unresolved.

Commenting on these events, Ron Herring called it 'a fascinating reincarnation' of another rally of the same farmers' organisation, Shetkari Sanghatana, which took place in 2001 at the Nilkaneshwar temple on the 'sacred Narmada river'. Its president, Sharad Joshi, addressed the crowd that had gathered to protest the government's decision to 'outlaw Navbharat 151, the first and perhaps best, early *Bt* cotton hybrid'. 'Over our dead bodies' the outcome read. 'They will have to walk over our corpses to destroy this crop. This is our satyagraha.' As Herring commented further, 'the Nilkaneshwar temple is where the blue-throated god Shiva swallowed poison to protect humanity, just as the *Bt* crops, these farmers claimed, saved rural people from sprayed pesticides that were poisoning them and their soil – and bankrupting them as well' (Ronald Herring, pers. comm., 2019; Roy *et al.* 2007).

And what about Bangladesh? Event EE-1 was provided by Mahyco to the Bangladesh Agricultural Research Institute (BARI), which introgressed the event into nine local eggplant lines. Efficacy trials were conducted, beginning in 2005 and continuing today. The government granted approval for the release of four of the resultant varieties for 'limited cultivation' in 2013 and seedlings were distributed to 20 farmers in four districts. In 2016 and 2017, BARI distributed seeds to ~13 000 farmers and in 2018 seeds were sold through the Bangladesh Agricultural Development Corporation (BADC) to an additional 17 950 farmers. This represents an acceptance rate of ~17% of all eggplant farmers in the country (Shelton *et al.* 2018).

What were the reasons for this success in Bangladesh, bearing in mind that this was the first GM crop released for cultivation in that country? One of the reasons must surely be the partnership that developed between Mahyco, USAID, Sathguru Managements Consultants (an organisation specialising in traditional consulting services as well as technology transfer and innovation advice) and Cornell University. It is now called the South Asia Eggplant Improvement Partnership (SAEIP) and it designated BARI as the lead institute to both produce and distribute *Bt* eggplants to farmers. Together SAEIP and BARI ensured stewardship of the crop in both the pre- and post-launch phases. Together these organisations succeeded in producing adequate amounts of high-quality seed to meet grower demand (Shelton *et al.* 2018).

Of note is that the four *Bt* eggplant lines that have been released are not hybrids, so growers can save seed. Inclusion of the EE1 event in a hybrid background would further increase the yield potential of *Bt* eggplant. Although BARI was originally the sole provider of the seed, it has since brought in BADC to further increase the amount of seed available (Shelton *et al.* 2018).

Another reason for the crop's success in Bangladesh is the successful refuge requirement that was set early on by the SAEIP at 5% to prevent the build-up of resistance to *Bt* in the EFSB population. But possibly the lynchpin for the sustainable production of *Bt* eggplants was the farmer training provided by BARI, and more recently by the Department of Agricultural Extension and the Agricultural Information

Service, to hundreds of farmers in dozens of districts. These efforts have the full support of the SAEIP.

But what about the opposition anti-GMO lobby? Despite considerable negative press coverage in the early days of the roll-out of the crop, the satisfaction shown by farmers has helped to win them the support of the government. The SAEIP has played a major role in this by providing the press with factual information, which has included the studies in the Philippines that showed nearly 100% control of EFSB by *Bt* eggplant (Hautea *et al.* 2016) and the fact that there are no negative effects on non-target arthropods (Navasero *et al.* 2016).

In addition, there is government support for the development of *Bt* eggplants. In the words of the Honourable Agriculture Minister, Begum Matia Chowdhury, spoken at a 2017 workshop, 'Development of brinjal fruit and shoot insect resistant-*Bt* brinjal is a success story of local and foreign collaboration. We will be guided by the science-based information, not by the non-scientific whispering of a section of people. Good science will move on its own course keeping the anti-science people down. As human beings, it is our moral obligation that all people in our country should get food and not go to bed on an empty stomach. Biotechnology can play an important role in this effort.'

In Bangladesh, as everywhere else in the developing world, government support is essential if farmers are to be allowed to test, and if proven effective, to grow GM crops.

Golden Rice

Rice is a staple crop for many, especially in Asia where it is eaten almost every day. In some countries 70–80% of an individual's calorie intake is from rice (Bouis and Islam 2012). However, during the removal of the husk and aleurone layer to prevent the grains from becoming rancid during storage, micronutrients are removed. In the case of rice, the most important one that is lost is vitamin A, leading to vitamin A deficiency (VAD). This has become a public health problem for many years and it is estimated that every year it kills 670 000 children under the age of 5 years (Black *et al.* 2008) and causes an additional 500 000 cases of irreversible childhood blindness (Humphrey *et al.* 1992).

In order to address this problem, two scientists, Ingo Potrykus and Peter Beyer, from the Institute of Plant Sciences at the Swiss Federal Institute of Technology (ETH) in Zurich, began work in the early 1990s to develop a variety of rice containing a source of vitamin A in its endosperm. The rice had to be created by introducing two new genes, one from maize and the other from a very commonly ingested soil bacterium, *Erwinia uredovora*. The product of these genes working together results in the formation of lycopene, which is converted by the plant into β-carotene. This compound has a red–orange pigment, which gave the resultant rice a golden colour. Once ingested, the body converts β-carotene into vitamin A.

In 2001, the inventors assigned their patents to Syngenta for commercialisation. Part of this transaction obliged the company to assist the scientists' humanitarian and altruistic objectives (Potrykus 2010). The terms included that there would be no charge for the nutritional technology and it would only be introduced into publicly owned rice varieties. In addition, there would be no limitations on what smallholder farmers could do with their crop. They can save and replant seed, sell seed and sell the final product (Dubock 2019). Improvements were made by scientists working for Syngenta and the International Institute for Rice Research (IRRI), which increased the effectiveness of Golden Rice (Paine *et al.* 2005). By then a Golden Rice Humanitarian Board had been formed to ensure, among other aspects, that, as Golden Rice is a not-for-profit project, no individual or organisation could have any financial interest in the development of the product.

In 2004 Syngenta ceased its commercial interest in Golden Rice and further development was funded only by philanthropic and public sectors. These included the governments of Bangladesh, China, India, Indonesia, the Philippines and Vietnam, as well as the US National Institutes of Health, USAID, the Rockefeller Foundation and the Bill & Melinda Gates Foundation (Dubock 2019).

Golden Rice has been the subject of major attacks by anti-GMO activists, spearheaded by Greenpeace, for many years. Reading about this crop on their website (Greenpeace South-east Asia 2013) it appears that their complaints are mainly about three issues. First, they believe that: 'by combating VAD with ecologically farmed home and community gardens, sustainable systems are created that provide food security and diversity in a way that is empowering people, protects biodiversity and ensures a long-lasting solution to VAD and malnutrition.' Second, they are of the opinion that: 'Golden rice is highly likely to contaminate non-GE rice, if released to the environment.' This will, they state: 'affect traditional, conventional and organic rice farmers because they will lose their markets, especially export markets, which would negatively impact rural livelihoods. If any hazardous, unexpected effects would develop from GE 'Golden' rice, the GE contamination would affect countries where rice is an essential staple and put people and food security at risk.' And, finally, they say that it is irresponsible to impose 'Golden' rice on people who oppose it on religious, cultural or personal grounds'.

On the first issue, no-one is against any way to combat VAD and if 'ecologically farmed home and community gardens' work, all strength to them. But, so far, none has been forthcoming. On the second one, scientists at the VIB (Vlaams Instituut voor Biotechnologie) in Belgium have studied Golden Rice's potential to cross-pollinate other rice varieties and found this to be limited because rice is typically self-pollinated. This work agrees with research carried out at IRRI that demonstrated that the chance of 'gene flow' is very low for the same reason. Moreover, rice pollen is only viable for 3–5 minutes. Finally, preventive measures such as staggered flowering dates and observing recommended distances to other rice fields could further limit this risk. On the third issue, no-one is forcing people to plant or eat Golden Rice, so if they oppose it on the grounds mentioned by Greenpeace, they are welcome to do so.

Incensed by this opposition, some 150 Nobel Laureates wrote an open letter to the leaders of Greenpeace, as well as to the United Nations and governments around the world (Roberts 2018) urging 'Greenpeace and its supporters to re-examine the experience of farmers and consumers worldwide with crops and foods improved through biotechnology ... and abandon their campaign against 'GMOs' in general and Golden Rice in particular.'

In an interview with Ed Regis, the author of *Golden Rice: The Imperiled Birth of a GMO Superfood*, Cameron English asks, 'There are a lot of transgenic crops being developed, so why did Golden Rice become such a lightning rod for controversy in the GMO debate?' The answer was, 'Because if it gets approved, works and ends up saving lives and sight, it will lead to greater acceptance of GMO foods in general, which is the very last thing that GMO opponents want. That cannot be said of any other GMO' (English and Regis 2019).

Finally, Golden Rice can at last be tested because on 10 December 2019, the Philippines' government authorised the use of variety GR2E in food, feed and for processing. In doing so the country became the first to approve this crop in a region where rice is the staple crop and VAD is a major health problem (Dubock *et al.* 2019). They join Australia, New Zealand, Canada and the USA in this approval, but these countries hardly need this technology. In the words of the Director General of IRRI, Matthew Morell, 'Each regulatory application that Golden Rice completes with national regulatory agencies takes us one step closer to bringing Golden Rice to the people who need it the most. The rigorous safety standards observed by the USFDA and other agencies provide a model for decision-making in all countries wishing to reap the benefits of Golden Rice' (Lynas 2018b). At last there is a chance that his hopes, and those of the many scientists behind this development, could become a reality.

Maize in Mexico

In December 2018 Andrés Manuel Lopez Obrador (AMLO) became President of Mexico, which appears to have given a boost to international activist groups and their local affiliates who wish to make Mexico a GMO-free country. Their focus is mostly on corn because this plant originated in Mexico, which is still home to 64 different maize varieties. The AMLO administration appointed some of the local activists to important government positions where they have been lobbying to maintain the current ban on GM maize. They are also trying to change Mexico's biosafety law to ban all GM crops (Ventura 2019).

This is nothing new in Mexico where international groups such as the Union of Concerned Scientists, headquartered in Cambridge,

Massachusetts, the Organic Consumers Association (based in Minnesota) and Greenpeace have been active since the early 2000s. Opposition was initiated when a team of scientists reported in 2001 the presence of transgenes in landraces of maize in Oaxaca state in southern Mexico. These results were tested by a different scientific team from the National Institute of Ecology who, in 2005, found no evidence of transgene movement into maize landraces in the same region (Ortiz-García *et al.* 2005).

Despite these findings, an anti-GMO advocacy group called the Union of Scientists Committed to Society (UCCS) was formed in 2006, followed the next year by Semillas de Vida (Seeds of Life). These groups, together with other non-governmental organisations (NGOs) and farmers' associations, formed a coalition to file a legal class action with the government to prohibit the release of GM maize. They won. Since then, over the last 12 years, GM maize imports have increased by 136% (Ávila 2019). This situation is unlikely to change, because the founder of UCCS, Elena Alvarez-Buylla, is the current science minister and primary science adviser to the president (Ventura 2019).

Papayas in Hawaii

Hawaii is hardly a developing country, but this story is illustrative of the damage that a group of anti-GMO critics can do to a recovering industry. In 1995 the Hawaiian papaya industry faced potential destruction by papaya ringspot virus (PRSV), which had been discovered in the Puna district of the island of Hawai'i where 95% of the papaya was grown (Gonsalves 1998). A research group led by Dennis Gonsalves from Cornell University developed a GM variety that was resistant to the PRSV and commercially released in 1998. The study was written up by APSnet as 'Transgenic virus resistant papaya: new hope for control of Papaya ringspot virus in Hawaii'. Soon the resistant variety, called the Rainbow papaya, dominated Hawaii's papaya exports, but instead of ending the storm, as the name might suggest, the crop unleashed its own tempest (Brodwin 2017).

In 2013 a Maui County Council member introduced a bill to ban GMOs from that island. By doing this the councillor hoped to make

Maui a model for the rest of the world. Anti-GMO activists were video-conferenced into the hearing to support the ban. Despite protests from farmers the bill was signed into law in 2014, but was removed from the legislation in 2015 when a federal judge ruled that Hawaiian counties could not enact their own GMO bans (Brodwin 2017).

As of 2018, Rainbow papaya counts for 75% of the crop but the issue is still being hotly contended (Auld 2018). Interestingly, while Hawaii debates these issues, China went ahead as early as 1996 to allow the commercial planting of virus-resistant papaya and in 2014 there was an adoption rate of between 93% and 94% (Li *et al*. 2014). Perhaps Hawaii could take a leaf out of China's book?

Genome editing

This section on genome editing is placed here because, as will be seen, this is a technology that could really make a difference to the lives of both the poor and the not-so-poor, but already the EU is threatening to stigmatise it as unsafe.

The first, and still widely used, method for genome editing is called CRISPR. It allows scientists to change DNA sequences in a very precise manner and thus change the function of a gene. It has the capacity to improve crops by, for instance, changing a gene that might code for sensitivity to a harmful pathogen.

How does it work? CRISPR is an acronym for Clustered Regularly Interspaced Short Palindromic Repeats. These short palindromic repeat sequences of ~30 base pairs are found clustered at regular intervals in a genome. CRISPR is itself shorthand for CRISPR/Cas9 (Crisper associated). The CRISPR part is a sequence of DNA that binds to the repeat sequence in a particular gene; for example, one that codes for sensitivity to a plant virus. CRISPR is bound to Cas-9, which is an enzyme that cuts the plant's DNA at that particular spot like a pair of scissors. The plant's own DNA repair system then comes into play to mend the cut by introducing random bases. Thus, a change or mutation is introduced at that spot and the gene no longer codes for virus sensitivity.

The technology, which was discovered and developed in 2012, has already been used for crop improvement; for instance, resistance in cassava to its brown streak virus, a major problem in East and Central Africa (Gomez *et al.* 2019). But how will regulators react to this new tool? One problem they face is that, unlike 'old fashioned' genetic engineering, the plant is left without any signature DNA sequences that can be used to detect standard GMOs. The only way to detect the difference between a conventional plant and its gene-edited counterpart is by fine-scale DNA sequencing to determine whether a particular gene has been mutated or not. Not a straight-forward exercise.

The first country to take the plunge was the USA, which gave the green light to the CRISPR-edited white mushroom that is resistant to browning. It was developed by a plant pathologist, Yinong Yang, at Pennsylvania State University, who deleted a few base pairs in a family of genes coding for the enzyme polyphenol oxidase that is responsible for browning (Waltz 2016).

In response to this, there is a rather interesting dialogue taking place in the organic farming sector. Most of them, exemplified by the pro-organic Cornucopia Institute in Wisconsin, USA, express their abhorrence for genetic engineering, including gene-edited crops. However, some organic growers see this technology as a boon for their industry and are calling for a revision in the rules so that they can benefit from gene editing. Among these is the influential Klaas Martens, who farms in New York's Finger Lakes region and runs Lakeview Organic Grain. At a conference in 2019 he said, 'If it could be used in a way that enhances the natural system, and mimicked it, then I would want to use it. But it would definitely have to be case by case.' Others noted that CRISPR-like techniques could help them to reduce pesticide use and grow more disease-resistant crops (Cerier 2019).

Moving on to Canada, we find that this country has remained committed to the scientific principles that it specified in 1992 when it first considered regulations for plants with novel traits. Since then the country has delivered timely and consistent approval decisions. Although the approval of the non-browning Arctic® apple took some time, the more recent approval in 2016 of the Innate® potato with decreased levels of asparagine and therefore reduction of potentially toxic acrylamide on cooking at high temperatures occurred more rapidly (Smyth 2017).

What about the EU? On 25 July 2018 the Court of Justice of the European Union determined that 'Organisms obtained by mutagenesis are GMOs and are, in principle, subject to the obligations laid down by the GMO directive.' However, 'organisms obtained by mutagenesis techniques which have conventionally been used in a number of applications and have a long safety record are exempt from these obligations' (Court of Justice of the European Union 2018).

The next day, in an article entitled 'Scientific community defeated by green groups in European court ruling on gene edited crops', Mark Lynas commented that Europe now faced the bizarre situation of regulating new plant varieties that had been produced by precise genome editing but allowing the 'hit-and-miss' mutagenesis by chemicals or radiation (Lynas 2018c).

In a response to this ruling, scientists representing 93 European plant and life sciences research centres and institutes endorsed a position paper calling upon the court to alter the legislation and to thoroughly revise the GMO regulations to 'correctly reflect scientific progress in biotechnology'. This is because 'organisms that have undergone simple and targeted genome edits by means of precision breeding and which do not contain foreign genes are at least as safe as if they were derived from classical breeding techniques' (VIB 2018).

The scientists expressed great concern that this ruling would put European agricultural industries at a competitive disadvantage because it would make precision breeding hyper-expensive. Farmers too would miss out on new crops that are more nutritious and better suited to respond to climate change (VIB 2018).

Subsequently, in May 2019, 14 EU countries called for a 'unified approach' to gene editing in plants. In response the European Commission promised to come up with a 'robust response' to the EU court ruling and draft a legislative proposal 'in due time' – whenever that may be (Fortuna 2019).

As far as developing countries are concerned, many of them could follow the lead of the EU, for the reasons outlined in Chapter 9, denying their farmers and seed companies access to these new crops to be bred with highly precise, modern technology. Time will tell whether science or politics holds sway.

References

Alliance for Science (2019) 'Indian farmers plant GMO seeds in civil disobedience 'satyagraha' protest'. 10 June 2019. <https://allianceforscience. cornell.edu/blog/2019/06/indian-farmers-plant-gmo-seeds-civil-disobedience-satyagraha-protest/>

Auld E (2018) 'GMOs in the Aloha State: biotech's passion for Hawaii'. *Living Non-GMO*, 18 October 2018. <https://livingnongmo.org/2018/10/18/gmos-in-the-aloha-state-biotechs-passion-for-hawaii/>

Ávila O (2019) 'Mexican corn, under threat; import increased 136% in the last ten years'. *Excelsior*, 30 September 2019. <https://translate.google.com/translate?hl=&sl=es&tl=en&u=https://www.excelsior.com.mx/nacional/el-maiz-mexicano-bajo-amenaza-importacion-aumento-136-en-los-ultimos-diez-anos/1339135>

Black RE, Allen LH, Bhutta ZA, Caulfield LE, de Onis M, Ezzati M, Maathers C, Rivera J (2008) Maternal and child undernutrition: global and regional exposures and health consequences. *Lancet* **371**, 243–260. doi:10.1016/S0140-6736(07)61690-0

Bouis H, Islam Y (2012) Biofortification: leveraging agriculture to reduce hidden hunger. In *Reshaping Agriculture for Nutrition and Health*. (Eds S Fan and R Pandya-Lorch) pp. 83–92. International Food Policy Research Institute, Washington, DC. <https://books.google.co.za/books?hl=en&lr=&id=gVQZ Xr38z5sC&oi=fnd&pg=PA83&dq=Bouis+H+and+Islam+Y+(2012)+Biofortification:+Leveraging+agriculture+to+reduce+hidden+hunger&ots=0Dg5ZY4nr 8&sig=Tz4z6Z0ILtGxjBgtWPIdRDjXpQk#v=onepage&q&f=false>

Brodwin E (2017) 'This Cornell scientist saved an $11-million industry – and ignited the GMO wars'. Business Insider, 24 June 2017. <https://www. businessinsider.com/gmo-controversy-beginning-fruit-2017-6?IR=T>

Cerier S (2019) 'Is there a place for CRISPR gene editing in organic farming? Many farmers say 'yes''. Genetic Literacy Project, 3 December 2019. <https://geneticliteracyproject.org/2019/12/03/is-there-a-place-for-crispr-gene-editing-in-organic-farming-many-farmers-say-yes/>

Conrow J (2016) 'Philippines supreme court reverses GMO ruling'. Alliance for Science, 26 July 2016. <https://allianceforscience.cornell.edu/blog/2016/07/philippines-supreme-court-reverses-gmo-ruling/>

Court of Justice of the European Union (2018) Judgment in Case C-528/16. <https://curia.europa.eu/jcms/upload/docs/application/pdf/2018-07/cp180111en.pdf>

Del Prado-Lu JL (2015) Insecticide residues in soil, water, and eggplant fruits and farmers' health effects due to exposure to pesticides. *Environmental Health and Preventive Medicine* **20**, 53–62. doi:10.1007/s12199-014-0425-3

Dubock A (2019) 'Golden Rice: to combat vitamin A deficiency for public health'. IntechOpen, 11 March 2019. <https://www.intechopen.com/books/vitamin-a/golden-rice-to-combat-vitamin-a-deficiency-for-public-health>

Dubock A, Potrykus I, Beyer P (2019) 'Philippines is first! Long-delayed Vitamin A-enhanced Golden Rice greenlighted, bucking activist opposition'. Genetic Literacy Project, 18 December 2019. <https://geneticliteracyproject.org/2019/12/18/philippines-is-first-long-delayed-vitamin-a-enhanced-golden-rice-greenlighted-bucking-activist-opposition/>

English C, Regis E (2019) 'How misguided regulation has kept a GMO 'superfood' off the market: Q&A with Golden Rice author Ed Regis'. Genetic Literacy Project, 5 November 2019. <https://geneticliteracyproject.org/2019/11/05/how-misguided-regulation-has-kept-a-gmo-superfood-off-the-market-qa-with-golden-rice-author-ed-regis/>

Fortuna G (2019) '14 EU countries call for 'unified approach' to gene editing in plants'. EURACTIV, 24 May 2019. <https://www.euractiv.com/section/agriculture-food/news/14-eu-countries-call-for-unified-approach-to-gene-editing-in-plants/>

Gomez MA, Lin ZD, Moll T, Chauhan RD, Hayden L, Renninger K, Beyene G, Taylor NG, Carrington JC, Staskawicz BJ, Bart RS (2019) Simultaneous CRISPR/Cas9-mediated editing of cassava *eIF4E* isoforms *nCBP-1* and *nCBP-2* reduces cassava brown streak disease symptom severity and incidence. *Plant Biotechnology Journal* **17**, 421–434. doi:10.1111/pbi.12987

Gonsalves D (1998) Control of papaya ringspot virus in papaya: a case study. *Annual Review of Phytopathology* **36**, 415–437. doi:10.1146/annurev.phyto.36.1.415

Greenpeace India (2009) 'Effects on health and environment of transgenic (or GM) Bt brinjal'. 17 April 2009. <https://www.greenpeace.org/india/en/publication/855/effects-on-health-and-environment-of-transgenic-or-gm-bt-brinjal/>

Greenpeace Southeast Asia (2013) 'Golden Rice'. <https://www.greenpeace.org/southeastasia/publication/1073/golden-rice/>

Guntur PSJ (2019) 'Farmers at their wits' end over fake seeds'. *The Hindu*, 29 June 2019. <https://www.thehindu.com/news/cities/Vijayawada/farmers-at-their-wits-end-over-fake-seeds/article28229245.ece>

Hautea DM, Taylo LD, Masanga APL, Sison MLJ, Narciso JO, Quilloy RB, Hautea RA, Shotkoski FA, Shelton AM (2016) Field performance of Bt eggplants (*Solanum melongena* L.) in the Philippines: cry1ac expression and control of the eggplant fruit and shoot borer (*Leucinodes orbonalis* Guenée). *PLoS One* **11**, e0157498. doi:10.1371/journal.pone.0157498

Herring RJ (2015) State science, risk and agricultural biotechnology: *Bt* cotton to *Bt* Brinjal in India. *The Journal of Peasant Studies* **42**, 159–186. doi:10.1080/03066150.2014.951835

Humphrey JH, West KP Jr, Sommer A (1992) Vitamin A deficiency and attributable mortality in under-5-year-olds. *Bulletin of the World Health Organization* **70**, 225–232.

Kouser S, Qaim M (2014a) Bt cotton, damage control and optimal levels of pesticide use in Pakistan. *Environment and Development Economics* **19**, 704–723. doi:10.1017/S1355770X1300051X

Kouser AS, Qaim M (2014b) Bt cotton, pesticide use and environmental efficiency in Pakistan. *Journal of Agricultural Economics* **66**, 66–86. doi:10.1111/1477-9552.12072

Kouser AS, Qaim M (2019) Transgenic cotton and farmers' health in Pakistan. *PLoS One* **14**, e0222617. doi:10.1371/journal.pone.0222617

Kulkarni P (2018) 'As a genetic revolution collapses, Vidarbha's cotton farmers dread coming season'. Bloomberg *Quint*, 21 July 2018. <https://www.bloombergquint.com/global-economics/as-a-genetic-revolution-collapses-vidarbhas-cotton-farmers-dread-coming-season>

Li Y, Peng Y, Hallerman EM, Wu K (2014) Biosafety management and commercial use of genetically modified crops in China. *Plant Cell Reports* **33**, 565–573. doi:10.1007/s00299-014-1567-x

Lynas M (2018a) 'The complicated truth behind GM cotton in India'. Alliance for Science, 2 August 2018. <https://allianceforscience.cornell.edu/blog/2018/08/complicated-truth-behind-gmo-cotton-india/>

Lynas M (2018b) 'US FDA approves Golden Rice'. Alliance for Science, 25 May 2018. <https://allianceforscience.cornell.edu/blog/2018/05/us-fda-approves-golden-rice/>

Lynas M (2018c) 'Scientific community defeated by green groups in European court ruling on gene edited crops'. Alliance for Science, 26 July 2018. <https://allianceforscience.cornell.edu/blog/2018/07/scientific-community-defeated-green-groups-european-court-ruling-gene-edited-crops/>

Navasero MV, Canadano RN, Hautea DM, Hautea RA, Shotkoski FA, Shelton AM (2016) Assessing potential impact of Bt eggplants on non-target arthropods in the Philippines. *PLoS One* **11**, e0165190. doi:10.1371/journal.pone.0165190

Ortiz-García S, Ezcurra E, Schoel B, Acevedo F, Soberón J, Snow AA (2005) Absence of detectable transgenes in local landraces of maize in Oaxaca, Mexico (2003–2004). *Proceedings of the National Academy of Sciences of the United States of America* **102**, 12338–12343. doi:10.1073/pnas.0503356102

Paine J, Shipton C, Chaggar S, Howells R, Kennedy M, Vernon G, Wright S, Hinchliffe E, Adams J, Silverstone A, Drake R (2005) Improving the nutritional value of Golden Rice through increased pro-vitamin A content. *Nature Biotechnology* **23**, 482–487. doi:10.1038/nbt1082

Potrykus I (2010) The private sector's role for public sector genetically engineered crop projects. *New Biotechnology* **30**, 578–581. doi:10.1016/j.nbt.2010.07.006

Roberts RJ (2018) The Nobel Laureates' campaign supporting GMOs. *Journal of Innovation & Knowledge* **3**, 61–65. doi:10.1016/j.jik.2017.12.006

Roy D, Herring RJ, Geisler CC (2007) Naturalising transgenics: official seeds, loose seeds and risk in the decision matrix of Gujarati cotton farmers. *The Journal of Development Studies* **43**, 158–176. doi:10.1080/00220380601055635

Shelton AM, Hossain MJ, Paranjape V, Azad AK, Rahman ML, Khan ASMMR, Prodhan MZH, Rashid MA, Majumder R, Hossain MA, Hussain SS, Huesing JE, McCandless L (2018) Bt eggplant project in Bangladesh: history, present status, and future direction. *Frontiers in Bioengineering and Biotechnology* **6**, 106. doi:10.3389/fbioe.2018.00106

Smyth SJ (2017) Canadian regulatory perspectives on genome engineered crops. *GM Crops and Food: Biotechnology in Agriculture and the Food Chain* **8**, 35–43. doi:10.1080/21645698.2016.1257468

Sushma M (2019) 'Damning allegations: India's fields still have *Bt* brinjal'. DownToEarth, 25 April 2019. <https://www.downtoearth.org.in/news/agriculture/damning-allegations-india-s-fields-still-have-bt-brinjal-64187>

The Tribune (2018) 'Dealer booked for selling fake Bt cotton seed'. *The Tribune*, 8 March 2018. <https://www.tribuneindia.com/news/punjab/dealer-booked-for-selling-fake-bt-cotton-seed/554523.html>

Ventura L (2019) 'Viewpoint: How international anti-biotech activists manipulate year-old Mexican government to block crop GMO innovations'. Genetic Literacy Project, 2 December 2019. <https://geneticliteracyproject.org/2019/12/02/viewpoint-how-international-anti-biotech-activists-manipulate-year-old-mexican-government-to-block-crop-gmo-innovations/?mc_cid=265bc8fd15&mc_eid=832c481596>

VIB (2018) *Regulating genome edited organisms as GMOs has negative consequences for agriculture, society and economy.* Position Paper on the ECJ ruling on CRISPR, 12 Nov 2018. <http://www.vib.be/en/news/Documents/Position%20paper%20on%20the%20ECJ%20ruling%20on%20CRISPR%2012%20Nov%202018.pdf>

Waltz E (2016) Gene-edited CRISPR mushroom escapes US regulation. *Nature* **532**, 293–303. doi:10.1038/nature.2016.19754

11

Conclusions

Why is it that people in the West, particularly those in the European Union, but also in the Nordic countries and parts of the USA, are so against GM crops? These crops have been on sale in the USA since 1996 and in many other countries for a considerable number of years. They include maize, soybean, sunflowers, alfalfa, potatoes, squash and papayas. Since then there has not been a single proven case of ill health related to their consumption by either humans or animals. This comes as no surprise to scientists, because every major regulatory body in the world has come to the conclusion that GM crops are as safe for human and animal consumption as conventional crops – organic or not. Why hasn't this evidence put a stop to the GMO controversy long ago?

One reason proposed by Marcel Kuntz from the Institut National de la Recherche Agronomique (National Institute of Agricultural Research) in France is that there is a postmodern assault on science in general and GM crops in particular (Kuntz 2012). Postmodern philosophy claims that there is no universal truth and that 'each social or political group should have the right to the reality that best suits them'. Kuntz goes on to say, 'the danger of a postmodern approach to science, that seeks to include all points of view as equally valid, is that it slows down or prevents much needed scientific research, even denying that science should have a role in such decisions.' He ends his article by saying, 'However, as the GMO dispute has shown, scientists will never be able to win in postmodern courtroom-style debates: all 'social constructs' of science are equal, but some are more equal than others.'

One reason for the continuation of the GMO debate could be the power of social media. For instance, in the case of the use of glyphosate, the Non-GMO Project, based in Washington State in the USA, recently tweeted that Mexico had joined countries including Austria,

Germany, Thailand and Vietnam in banning GMO imports (McElrath 2019). However, Cameron English, writing for the Genetic Literacy Project, said that this was misleading because Austria and Thailand have now lifted the ban (English 2019). But this really makes no difference, as few people reading this tweet would check its truth and hence the original post served its purpose.

We just have to look at the effect that the antivaccination social media has had on the incidence of measles worldwide. In a survey by the Royal Society for Public Health it was found that half of the interviewed British parents with young children regularly came across antivaccination messages on social media (Burki 2019). According to the World Health Organization report of 12 August 2019, 'there have been almost three times as many cases reported to date in 2019 as there were at this same time last year' (World Health Organization 2019).

Taking a look at the role that social media have in the perpetuation of pseudoscientific 'facts', an article recently asked the question: 'Why are the wealthy more likely to fall for food pseudoscience?' (Buhler and Kirshenbaum 2019). They quoted a comprehensive 2016 study undertaken by the US National Academy of Sciences that concluded that GM crops are just as safe to eat as their non-GM counterparts. However, in a survey conducted by these authors, 43% of their respondents in higher income brackets reported that they avoided purchasing them, compared with only 26% of lower earners. The authors suggested that affluent Americans were more likely to encounter 'pseudoscientific facts' – in other words, unsubstantiated information – whether online, from family and friends, at farmers' markets or at upscale grocery stores. Although higher incomes might allow more access to information that could shape attitudes about diet and nutrition, 'higher income does not consistently correlate with better understanding. We believe they show the need for food experts and health professionals to work with social scientists to understand ways in which different communities make decisions about food' (Buhler and Kirshenbaum 2019).

The bottom line as far as the subject of this book is concerned, is that if many organisations in 'developed countries' express anti-GMO sentiment, many countries in the 'developing world' will take note. We only have to look at the status of GM crops in Africa to see the effect of this. Only South Africa and Sudan have a history of growing GM crops and in Sudan it is only cotton, whose seeds are only edible if processed to remove the toxic gossypol. The fact that Nigeria has opened its doors to both *Bt* cotton and *Bt* cowpea could lead to others following. At least by allowing farmers to test GM crops such as these, they will be able to decide for themselves if they wish to continue to grow them.

Perhaps at this point it is worth remembering what I wrote in the Preface regarding the difference between the 'West' and the 'Rest' and between a 'developed' and a 'developing' country. As Ron Herring commented to me, 'let us hope that all countries are developing, otherwise they would surely be regressing!'

Whichever definition one uses, however, it remains important how a country's 'state science' operates in decision making regarding GM crops. In his discussion of this topic, Herring (2015) presents as an example how a small number of scientists, linked to industry, were able to create doubt about the science behind climate change. Obviously, the industries involved had a vested interest in the outcome of the discussion and, certainly in the USA, they achieved their desired goal. Because the amount of scientific information that an ordinary citizen would require to make her/his own opinion is daunting, people will tend to rely on the regulatory authorities in their country to make the right decisions. Very often the science that such a body relies on is not straight *science*, but 'science filtered, weighted and selected through political processes constituted by or aimed at the state. Variations along these dimensions help explain patterns of states' acceptance and rejection of agricultural biotechnology but are not always decisive: interests are at stake' (Herring 2015).

Perhaps one way to sum up the effects that the West is having on the Rest is that the wealthier 'haves' who do not need to have 'it' (GM crops), because they have enough food already, are preventing the 'have nots', who do not have enough, from improving their food security. Farmers in poor countries can often not afford modern technologies such as pesticides, fertilisers and mechanisation, but they can more easily afford improved seeds. As a result, they have overtaken the developed world in the amount of GM crops grown. In 2014, almost 53% of the nearly 2 million square kilometres of such crops being grown worldwide was grown in 'developing' countries (James 2014). For the West to prevent such farmers from even testing whether the improved seed will bring improvements into their lives, and the lives of consumers, is morally concerning.

And what about the future? Will Generation Z, born between 1995 and 2015, make the necessary changes? Will approval processes, which are so expensive that currently only big companies can afford to bring GM crops to market, be simplified to allow access by small organisations and even academics? The current environmental movements advocate lifestyle changes to combat climate change, but most of them still hold quaint notions of what is 'natural' in agriculture. Although it is undeniable that modern agriculture has significant environmental impacts, many studies have shown that GM and gene-edited crops can mitigate against these (ISAAA 2018). The answer, as espoused by Barnard in an article entitled 'Viewpoint: Generation Z to environmentalists: If you care about sustainability, embrace GMOs and gene edited crops' (Barnard 2019), is to 'embrace biotechnology as a means to not only sustain and feed millions of people more effectively, but also to overcome many of the environmental problems associated with intensive agriculture.' So maybe we should be putting our hopes in this new generation?

Looking to the future, will the public and regulators accept gene-editing methodologies such as CRISPR? As discussed in Chapter 6, trust is the primary requirement. Thus, the proponents of this technology, whether they are the scientists at the laboratory bench, the seed companies that test the crops, the farmers who grow them or the production and marketing companies that put the products on the store shelves, all of these must earn trust.

One thing we have learnt from the GMO debate is that simply by educating people about the technologies involved – giving them the facts – will not cause them to accept gene editing. Science alone cannot answer the public's concerns regarding benefits and risks. This was highlighted in an article that studied the influence of four messages on the public's acceptance of the scientific consensus on the safety of GMOs (Landrum *et al.* 2019). Two messages were positive and two were negative. The first positive one came from the US National Academies of Sciences, Engineering and Medicine (NASEM), which in effect stated (albeit very carefully worded) that their consensus was that GM crops are safe for the environment and human consumption. The second was the consensus letter from nearly 150 Nobel Prize winners, as mentioned in Chapter 10, chastising Greenpeace for its anti-GMO stand. The two opposing messages came from 300 members of the European Network of Scientists for Social and Environmental Responsibility (ENSSER), stating that there was no scientific consensus on GMO safety, and an article written by a scholar and a journalist that challenged the NASEM panelists' impartiality (Krimsky and Schwab 2017). The authors of the article studying the effects of these four messages asked whether the public differentiated among the competing voices and whether their attitudes were influenced by the messages. Alternatively, was the public primarily driven by their pre-existing views?

Of the Nobel Prize winners who were signatories to the letter, while all being highly rated experts in their own fields, very few are leaders in the field of agricultural biotechnology. This was even more relevant in the case of the extremely disparate ENSSER group. The results were that, at least in the short-term, 'appeals to experts and presentation of hard facts about the safety of GMOs do not appear to be effective in changing attitudes.' Even when participants concluded that pro-consensus messages were stronger and more likely to reflect the scientific community's attitudes, those messages did not diminish their concern regarding GMOs. The authors concluded that instead of relying on consensus messaging, a stronger role could be played by motivated reasoning.

Thus, if we can learn anything from the errors of the GMO controversies to avoid them in the upcoming gene-editing debate, there has to be true dialogue, where opportunities are given for all people with concerns to be listened to and taken seriously. There is a real danger of scientists providing the answers to questions that are not being asked. This is clearly spelt out by Carmen Bain in her article entitled 'Will the public accept gene-edited foods? A social science perspective' (Bain 2019).

One of the mistakes that Bain advises proponents of GM crops to guard against is the argument that gene editing is no different from classical, traditional plant breeding. This assumes that if the public understands plant breeding, they will automatically accept gene editing. This argument is clearly not true because traditional plant breeding does not require laboratory interventions. By the same token, breeding by mutagenesis also needs laboratory intervention and is clearly not the same as classical plant breeding – but it, unlike GM crops, does not require regulation. What gene editing does is either mimic changes that could happen by chance in nature (one or two base changes) or perform genetic engineering (GE) in a far more precise manner than could be done before with 'classic' GE technology.

We need to avoid reducing the debate to whether the public will accept gene editing and instead reflect on lessons from the GMO debate to consider what factors may enhance or impede public trust. This will be critical to building the credibility and trustworthiness of the organisations and experts involved.

One of the problems that over-zealous proponents of GM crops have left us with has been their propensity to view these crops as 'silver bullets' that can potentially cure all agricultural woes. And, of course, in the words of Ronald Herring, 'silver bullets are easy to knock down. Unfortunately, advertising imperatives of developers reinforce that narrative. There are no silver bullets, only useful interventions in very complex systems.' Take, for instance, the example of insects developing resistance against the *Bt* toxin. It is easy for anti-GMO activists to say that *Bt* cotton has failed. But in truth there is no such thing as *Bt* cotton, 'just hundreds of cultivars that share one trait in common: *Bt*. The trait does not the cultivar make' (Ronald Herring, pers. comm., 2019).

Let us now turn to an issue that still needs to be decided, the potential ban on the use of glyphosate (discussed in Chapter 6). Farmers in developing countries who will bear the heaviest burden of such a ban are those who live in the top 10 GM crop-planting countries, such as Brazil, Argentina, India, Paraguay, South Africa and Uruguay (ISAAA 2018). They use this herbicide because it helps them to control weeds at much lower costs. Indeed, in India, the cost of manual weeding can be as much as three times higher than if the Roundup® spray is used (Varshney 2018).

The intense backlash against this herbicide is based on the International Agency for Research in Cancer's (IARC) 2015 decision that glyphosate is a 'probable carcinogen' despite, as discussed in Chapter 6, many expert individuals and organisations disagreeing with this judgement on the grounds, among others, that the IARC did not take exposure and risk into account.

In trying to answer this conundrum, Kabat (2019) quotes a Nobel Prize-winning behavioural psychologist, Daniel Kahneman, as calling this an example of an 'availability cascade ... a self-sustaining chain of events, which may start from media reports of a relatively minor event and lead up to public panic and large-scale government action.' In addition, Kahneman considers that 'the danger is increasingly exaggerated as the media compete for attention-grabbing headlines. Scientists and others who try to dampen the increasing fear and revulsion attract little attention, most of it hostile: anyone who claims that the danger is overstated is suspected of association with a 'heinous cover-up.'

Ending his article, Kabat (2019) states that these are complex issues that create a 'near-impenetrable thicket of uncertainties', but, in this case the 'clear weight of evidence coupled with a dose of common sense is enough to show what's right ... Glyphosate is a boon to agriculture and humanity. Let's refocus the energy and resources spent on trying to demonize this useful and valuable chemical on problems that really matter.'

Unless we do just that, one of the countries that would suffer most from a glyphosate ban is Argentina. As mentioned in Chapter 5, farmers in that country not only received the usual advantages of using this biodegradable herbicide that others do, but in addition the 'no tilling' method of farming means that they can obtain a 'second crop' of soybeans after wheat in the same season. In their article entitled 'The contribution of glyphosate to agriculture and potential impact of restrictions on use at the global level' (Brookes *et al.* 2017), the authors calculated that Argentina would lose US$1.4 billion annually from such a ban (21.2% of the global annual loss of US$6.76 billion). Brazil will not be far behind, losing US$818 million. In both these countries the losses will derive mostly from soybeans and maize.

The authors go on to state that these production losses in Argentina would likely result in a significant reduction in the 'second' soybean crop. They estimate that this could account for over 85% of that country's total soybean production, as 'farmers would be expected, in the long run to switch away from reduced/no tillage production systems to conventional tillage (due to greater difficulties in obtaining good weed control without access to glyphosate), leading to a longer growing season for soybeans and hence fewer to plant a soybean crop after wheat in the same season.'

So yet again, decisions in the West could have a significant negative effect on the livelihoods of those living in the rest of the world. And, yet again, the action is based on a decision of one organisation that goes against those of many others. Does that mean the instigators of the glyphosate ban are 'anti-science'? In an article entitled '(Practically) no one is anti-science, and how can that help us talk about GMOs', Marc Brazeau (2019) argues that the charge is rooted in the word science being used imprecisely. In his view, the term is more correctly 'anti-scientific method'.

In the case of GMOs, scientific answers cannot be found by constructing just one experiment to prove, or disprove, a hypothesis such as 'Do GM crops have harmful effects on the environment?' As this cannot be answered by a single study, the entire literature on the subject needs to be studied. And here, as Brazeau points out, lies the rub. Outlier studies can be held to have greater importance than the sum of all the others. In his words: 'to reject the weight of evidence based on one or a few outliers makes no sense and is not science.' He gives as an example the case for or against Golden Rice. Instead of asking the question 'Why do you want children to go blind by opposing Golden Rice?' we should be asking, 'Why do you see a handful of poorly conducted studies as the signal and the hundreds if not thousands of well conducted studies summarized in literature reviews and meta-analyses as the noise?' As he points out, 'Following science means we do not cherry-pick studies that reflect our predetermined values.' Unfortunately, as has been shown time and again, plausible risk is confused with imaginary risk, often resulting in regulatory delays, one of the reasons for my writing this book.

Finally, let us turn to the other burning issue that still has to be resolved, that of gene editing using methods such as CRISPR/Cas9. The potential this technology has for crop improvement and food sustainability is enormous. Let us learn from the mistakes made over GM crops and not repeat them. In particular, let us be aware that decisions made in the West can have a huge impact on the actions taken by the Rest.

References

Bain C (2019) 'Will the public accept gene-edited foods? A social science perspective'. Seed World, 18 December 2019. <https://seedworld.com/will-the-public-accept-gene-edited-foods-a-social-science-perspective/>

Barnard C (2019) 'Viewpoint: Generation Z to environmentalists: if you care about sustainability, embrace GMOs and gene edited crops'. Genetic Literacy Project, 25 November 2019. <https://geneticliteracyproject.org/2019/11/25/viewpoint-generation-z-to-environmentalists-if-you-care-about-sustainability-embrace-gmos-and-gene-edited-crops/>

Brazeau M (2019) '(Practically) no one is anti-science, and how that can help us talk about GMOs'. Genetic Literacy Project, 12 July 2019. <https://geneticliteracyproject.org/2019/07/12/practically-no-one-is-anti-science-and-how-that-can-help-us-talk-about-gmos/>

Brookes G, Taheripour F, Tyner WE (2017) The contribution of glyphosate to agriculture and potential impact of restrictions on use at the global level. *GM Crops and Food: Biotechnology in Agriculture and the Food Chain* **8**, 216–228. doi:10.1080/21645698.2017.1390637

Buhler D, Kirshenbaum S (2019) 'From GMOs to BPA, why the wealthy are more likely to fall for food pseudoscience'. Genetic Literacy Project, 20 September 2019. <https://geneticliteracyproject.org/2019/09/20/from-gmos-to-bpa-why-are-the-wealthy-more-likely-to-fall-for-food-pseudoscience/>

Burki T (2019) Vaccine misinformation and social media. *The Lancet Digital Health* **1**, 258–259. doi:10.1016/S2589-7500(19)30136-0

English E (2019) 'Natural health and conspiracy sites exploit social media to fester opposition to GMO crops. Here's a study about what can be done to

stop it'. Genetic Literacy Project, 10 December 2019. <https://
geneticliteracyproject.org/2019/12/10/natural-health-and-conspiracy-sites-
exploit-social-media-to-fester-opposition-to-gmo-crops-heres-a-study-about-
what-can-be-done-to-stop-it/>

Herring RJ (2015) State science, risk and agricultural biotechnology: *Bt* cotton
to *Bt* Brinjal in India. *The Journal of Peasant Studies* **42**, 159–186.
doi:10.1080/03066150.2014.951835

ISAAA (2018) 'Global status of commercialized biotech/GM crops in 2018:
biotech crops continue to help meet the challenges of increased population
and climate change'. ISAAA Brief No. 54. ISAAA, Ithaca, NY.

James C (2014) 'Global status of commercialized biotech/GM crops: 2014'.
ISAAA Brief No. 49. ISAAA, Ithaca, NY.

Kabat G (2019) 'Viewpoint: how the glyphosate-cancer controversy became a
moral crusade – and a threat to scientific progress'. Genetic Literacy
Project, 16 December 2019. <https://geneticliteracyproject.org/2019/12/16/
viewpoint-how-the-glyphosate-cancer-controversy-became-a-moral-
crusade-and-a-threat-to-scientific-progress>

Krimsky S, Schwab T (2017) Conflicts of interest among committee members
in the national academies' genetically engineered crop study. *PLoS One* **12**,
e0172317. doi:10.1371/journal.pone.0172317

Kuntz M (2012) The postmodern assault on science. *EMBO Reports* **13**,
885–889. doi:10.1038/embor.2012.130

Landrum AR, Hallman WK, Jamieson KH (2019) Examining the impact of
expert voices: communicating the scientific consensus on genetically-
modified organisms. *Environmental Communication* **13**, 51–70.
doi:10.1080/17524032.2018.1502201

McElrath KJ (2019) 'Mexico joins growing list of nations banning glyphosate'.
The Ring of Fire, 26 November 2019. <https://trofire.com/2019/11/26/
mexico-joins-growing-list-of-nations-banning-glyphosate/>

Varshney V (2018) 'Glyphosate use increased 1500% since genetically modified
crops were introduced'. DownToEarth, 26 July 2018. <https://www.
downtoearth.org.in/blog/food/glyphosate-use-increased-1500-since-
genetically-modified-crops-were-introduced-61241>

World Health Organization (2019) 'New measles surveillance data from
WHO'. Provisional data based on monthly reports to WHO (Geneva) as of
August 2019. <https://www.who.int/immunization/newsroom/new-
measles-data-august-2019/en/>

Index